FLORIDA BAY RESEARCH PROGRAMS
AND THEIR RELATION TO THE
COMPREHENSIVE EVERGLADES RESTORATION PLAN

Committee on Restoration of the Greater Everglades Ecosystem

Water Science and Technology Board

Board on Environmental Studies and Toxicology

Division on Earth and Life Studies

NATIONAL RESEARCH COUNCIL
OF THE NATIONAL ACADEMIES

THE NATIONAL ACADEMIES PRESS
Washington, D.C.
www.nap.edu

THE NATIONAL ACADEMIES PRESS   500 Fifth Street, N.W.   Washington, DC 20001

NOTICE: The project that is the subject of this report was approved by the Governing Board of the National Research Council, whose members are drawn from the councils of the National Academy of Sciences, the National Academy of Engineering, and the Institute of Medicine. The members of the committee responsible for the report were chosen for their special competences and with regard for appropriate balance.

Supported by the South Florida Ecosystem Restoration Task Force, U.S. Department of the Interior, under assistance of Cooperative Agreement No. 5280-9-9029. The views and conclusions contained in this document are those of the authors and should not be interpreted as necessarily representing the official policies, either expressed or implied, of the U. S. Government.

International Standard Book Number 0-309-08491-1

Additional copies of this report are available from the National Academies Press, 500 Fifth Street, N.W., Lockbox 285, Washington, DC 20055; (800) 624-6242 or (202) 334-3313 (in the Washington metropolitan area); Internet, http://www.nap.edu

Copyright 2002 by the National Academy of Sciences. All rights reserved.

Printed in the United States of America

THE NATIONAL ACADEMIES
Advisers to the Nation on Science, Engineering, and Medicine

The **National Academy of Sciences** is a private, nonprofit, self-perpetuating society of distinguished scholars engaged in scientific and engineering research, dedicated to the furtherance of science and technology and to their use for the general welfare. Upon the authority of the charter granted to it by the Congress in 1863, the Academy has a mandate that requires it to advise the federal government on scientific and technical matters. Dr. Bruce M. Alberts is president of the National Academy of Sciences.

The **National Academy of Engineering** was established in 1964, under the charter of the National Academy of Sciences, as a parallel organization of outstanding engineers. It is autonomous in its administration and in the selection of its members, sharing with the National Academy of Sciences the responsibility for advising the federal government. The National Academy of Engineering also sponsors engineering programs aimed at meeting national needs, encourages education and research, and recognizes the superior achievements of engineers. Dr. Wm. A. Wulf is president of the National Academy of Engineering.

The **Institute of Medicine** was established in 1970 by the National Academy of Sciences to secure the services of eminent members of appropriate professions in the examination of policy matters pertaining to the health of the public. The Institute acts under the responsibility given to the National Academy of Sciences by its congressional charter to be an adviser to the federal government and, upon its own initiative, to identify issues of medical care, research, and education. Dr. Harvey V. Fineberg is president of the Institute of Medicine.

The **National Research Council** was organized by the National Academy of Sciences in 1916 to associate the broad community of science and technology with the Academy's purposes of furthering knowledge and advising the federal government. Functioning in accordance with general policies determined by the Academy, the Council has become the principal operating agency of both the National Academy of Sciences and the National Academy of Engineering in providing services to the government, the public, and the scientific and engineering communities. The Council is administered jointly by both Academies and the Institute of Medicine. Dr. Bruce M. Alberts and Dr. Wm. A. Wulf are chair and vice chair, respectively, of the National Research Council

www.national-academies.org

COMMITTEE ON RESTORATION OF THE GREATER EVERGLADES ECOSYSTEM (CROGEE)[1]

JEAN M. BAHR[2], *Chair*, University of Wisconsin, Madison
SCOTT W. NIXON[2], *Vice-Chair*, University of Rhode Island, Narragansett
JOHN S. ADAMS, University of Minnesota, Minneapolis
LINDA K. BLUM[2], University of Virginia, Charlottesville
PATRICK L. BREZONIK[2], University of Minnesota, St. Paul
FRANK W. DAVIS, University of California, Santa Barbara
WAYNE C. HUBER[2], Oregon State University, Corvallis
STEPHEN R. HUMPHREY, University of Florida, Gainesville
DANIEL P. LOUCKS, Cornell University, Ithaca, New York
KENNETH W. POTTER, University of Wisconsin, Madison
LARRY ROBINSON, Florida Agricultural and Mechanical University, Tallahassee
REBECCA R. SHARITZ, Savannah River Ecology Laboratory, Aiken, South Carolina, and University of Georgia, Athens
HENRY J. VAUX, JR. University of California Division of Agriculture and Natural Resources, Oakland
JOHN VECCHIOLI, U.S. Geological Survey (ret.), Odessa, Florida
JEFFREY R. WALTERS, Virginia Polytechnic Institute and State University, Blacksburg

NRC Staff

STEPHEN D. PARKER, Director, Water Science and Technology Board
DAVID J. POLICANSKY, Associate Director, Board on Environmental Studies and Toxicology
WILLIAM S. LOGAN[2], Senior Staff Officer, Water Science and Technology Board
PATRICIA JONES KERSHAW, Staff Associate, Water Science and Technology Board

---

[1] The activities of the Committee on Restoration of the Greater Everglades Ecosystem (CROGEE) are overseen and supported by the NRC's Water Science and Technology Board (lead) and Board on Environmental Studies and Toxicology.

[2] A subgroup consisting of CROGEE members Wayne Huber (subgroup chair), Linda Blum, Patrick Brezonik, Scott Nixon, and CROGEE chair Jean Bahr, with support by NRC senior staff officer William Logan, took the lead in drafting the report.

WATER SCIENCE AND TECHNOLOGY BOARD

RICHARD G. LUTHY, *Chair,* Stanford University, Stanford, California
JOAN B. ROSE, *Vice Chair,* University of South Florida, St. Petersburg
RICHELLE M. ALLEN-KING, Washington State University, Pullman
GREGORY B. BAECHER, University of Maryland, College Park
KENNETH R. BRADBURY, Wisconsin Geological and Natural History
    Survey, Madison
JAMES CROOK, CH2M Hill, Boston, Massachusetts
EFI FOUFOULA-GEORGIOU, University of Minnesota, Minneapolis
PETER GLEICK, Pacific Institute for Studies in Development,
    Environment, and Security, Oakland, California
JOHN LETEY, JR., University of California, Riverside
DIANE M. MCKNIGHT, University of Colorado, Boulder
CHRISTINE L. MOE, Emory University, Atlanta, Georgia
ROBERT PERCIASEPE, National Audubon Society, Washington, D.C.
RUTHERFORD H. PLATT, University of Massachusetts, Amherst
JERALD L. SCHNOOR, University of Iowa, Iowa City
LEONARD SHABMAN, Virginia Polytechnic Institute and State
    University, Blacksburg
R. RHODES TRUSSELL, Montgomery Watson, Pasadena, California

Staff

STEPHEN D. PARKER, Director
LAURA J. EHLERS, Senior Staff Officer
JEFFREY W. JACOBS, Senior Staff Officer
WILLIAM S. LOGAN, Senior Staff Officer
MARK C. GIBSON, Staff Officer
STEPHANIE E. JOHNSON, Consulting Staff Officer
M. JEANNE AQUILINO, Administrative Associate
ELLEN A. DE GUZMAN, Research Associate
PATRICIA JONES KERSHAW, Study/Research Associate
ANITA A. HALL, Administrative Assistant
ANIKE L. JOHNSON, Project Assistant
JON Q. SANDERS, Project Assistant

BOARD ON ENVIRONMENTAL STUDIES AND TOXICOLOGY

GORDON ORIANS *(Chair)*, University of Washington, Seattle
JOHN DOULL *(Vice Chair)*, University of Kansas Medical Center, Kansas City
DAVID ALLEN, University of Texas, Austin
INGRID C. BURKE, Colorado State University, Fort Collins
THOMAS BURKE, Johns Hopkins University, Baltimore, Maryland
WILLIAM L. CHAMEIDES, Georgia Institute of Technology, Atlanta
CHRISTOPHER B. FIELD, Carnegie Institute of Washington, Stanford, California
DANIEL S. GREENBAUM, Health Effects Institute, Cambridge, Massachusetts
BRUCE D. HAMMOCK, University of California, Davis
ROGENE HENDERSON, Lovelace Respiratory Research Institute, Albuquerque, New Mexico
CAROL HENRY, American Chemistry Council, Arlington, Virginia
ROBERT HUGGETT, Michigan State University, East Lansing
JAMES H. JOHNSON, Howard University, Washington, D.C.
JAMES F. KITCHELL, University of Wisconsin, Madison
DANIEL KREWSKI, University of Ottawa, Ottawa, Ontario
JAMES A. MACMAHON, Utah State University, Logan
WILLEM F. PASSCHIER, Health Council of the Netherlands, The Hague
ANN POWERS, Pace University School of Law, White Plains, New York
LOUISE M. RYAN, Harvard University, Boston, Massachusetts
KIRK SMITH, University of California, Berkeley
LISA SPEER, Natural Resources Defense Council, New York, New York

Staff

JAMES J. REISA, Director
DAVID J. POLICANSKY, Associate Director and Senior Program Director for Applied Ecology
RAYMOND A. WASSEL, Senior Program Director for Environmental Sciences and Engineering
KULBIR BAKSHI, Program Director for the Committee on Toxicology
ROBERTA M. WEDGE, Program Director for Risk Analysis
K. JOHN HOLMES, Senior Staff Officer
SUSAN N.J. MARTEL, Senior Staff Officer
SUZANNE VAN DRUNICK, Senior Staff Officer
RUTH E. CROSSGROVE, Managing Editor

Preface

This report is a product of the Committee on Restoration of the Greater Everglades Ecosystem (CROGEE; an acronym list is given in Appendix A), which provides consensus advice to the South Florida Ecosystem Restoration Task Force ("Task Force"). The Task Force was established in 1993 and was codified in the 1996 Water Resources Development Act (WRDA); its responsibilities include the development of a comprehensive plan for restoring, preserving and protecting the South Florida ecosystem, and the coordination of related research. The CROGEE works under the auspices of the Water Science and Technology Board and the Board on Environmental Studies and Toxicology of the National Research Council.

The CROGEE's mandate includes providing the Task Force not only with scientific overview and technical assessment of the restoration activities and plans, but also providing focused advice on technical topics of importance to the restoration efforts. One such topic was to examine "the linkage between the upstream components of the greater Everglades and adjacent coastal ecosystems." This report addresses this issue by breaking it down into three major questions:

- What is the present state of knowledge of Florida Bay ("the Bay") on scientific issues that relate to the success of the overall CERP?
- What are the potential long-term effects of Everglades restoration as currently designed on the nature and condition of the Bay?
- What are the critical science questions that should be answered early in the restoration process to design a system that benefits not only the terrestrial and freshwater aquatic Everglades but the Bay as well?

This study was inspired in part by the 2001 Florida Bay and Adjacent Marine Systems Science Conference held on April 23-26, 2001 in Key Largo, Florida. An overlapping meeting of the CROGEE was held at the same location on April 26-28, 2001. The conference was organized by the Program Management Committee (PMC) of the Florida Bay and Adjacent Marine Systems Science Program (http://www.aoml.noaa.gov/flbay/). The PMC organized the conference around five questions suggested by the Florida Bay Science Oversight Panel (http://www.aoml.noaa.gov/flbay/oversight_panel.html). These questions related to circulation, salinity patterns, and outflows of the Bay; nutrients and the nutrient budget; onset, persistence and fate of planktonic algal blooms; temporal and spatial changes in seagrasses and the hardbottom community; and recruitment, growth and survivorship of higher trophic level species (http://www.aoml.noaa.gov/ocd/sferpm/stratpla.html). Some of these issues are discussed in the present report. However, as noted earlier, this report focuses on the subset of questions that relate to linkages between the Bay and the upstream portion of the Everglades system that arose at the 2001 Florida Bay Conference. As such, many science issues of importance to the health of the Bay but not directly related to the CERP are not discussed here, nor is this report intended to be a comprehensive review of the literature.

The conference provided an excellent environment to engage Bay scientists formally and informally during poster sessions, "synthesis sessions," and other venues. This report is based on an analysis of information presented at the Florida Bay Conference and a review of pertinent peer-reviewed literature.

The CROGEE is grateful for the assistance of many individuals during the data collection phase of this report. These include Peter Ortner (NOAA), South Florida Ecosystem Restoration Working Group liaison to CROGEE; Terrence "Rock" Salt, Executive Director of the Task Force; members of the Program Management Committee and Florida Bay Science Oversight Panel of the Florida Bay and Adjacent Marine Systems Science Program; and the many scientists at the conference who freely shared their insights into the complex issues regarding the ecology, hydrology, and water quality of the Bay. I would also like to thank the CROGEE members for their work on this report, especially a subgroup led by Wayne Huber and including Linda Blum, Patrick Brezonik, and Scott Nixon who took the lead in drafting the report with assistance from NRC staff officer Will Logan.

This report has been reviewed in draft form by individuals chosen for their diverse perspectives and technical expertise, in accordance with procedures approved by the NRC's Report Review Committee. The purpose of this independent review is to provide candid and critical comments that will assist the institution in making its published report as sound as possible and to ensure that the report meets institutional standards for objectivity, evidence, and responsiveness to the study charge. The review comments and draft manuscript remain confidential to protect the integrity of the deliberative process.

We wish to thank the following individuals for their review of this report:

Donald F. Boesch, University of Maryland Center for Environmental Science
Christopher Field, Carnegie Institution of Washington
Mandy Joye, University of Georgia
W. Michael Kemp, University of Maryland Center for Environmental Science
Diane McKnight, University of Colorado
Len Pietrafesa, North Carolina State University
Joe Rudek, North Carolina Environmental Defense
Sybil Seitzinger, Intergovernmental Oceanographic Commission of UNESCO
Y. Peter Sheng, University of Florida
William Wise, University of Florida

Although the reviewers listed above have provided many constructive comments and suggestions, they were not asked to endorse the conclusions or recommendations nor did they see the final draft of the report before its release. The review of this report was overseen by Jerry Robert Schubel, President and CEO, Aquarium of the Pacific, Long Beach, California. Appointed by the National Research Council, Dr. Schubel was responsible for making certain that an independent examination of this report was carried out in accordance with institutional procedures and that all review comments were carefully considered. Responsibility for the final content of this report rests entirely with the authoring committee and the institution.

Jean M. Bahr, Chair
Committee on Restoration of the Greater Everglades Ecosystem

Contents

EXECUTIVE SUMMARY 1

1 INTRODUCTION 5

2 POTENTIAL EFFECT OF THE COMPREHENSIVE EVERGLADES RESTORATION PLAN (CERP) ON FLORIDA BAY 9

3 RESEARCH NEEDS 17

4 CONCLUSIONS AND RECOMMENDATIONS 25

REFERENCES 29

APPENDIX A
    ACRONYM LIST 35

APPENDIX B
    DEFINITIONS OF MODEL RUNS OF THE SOUTH FLORIDA WATER MANAGEMENT MODEL (SFWMM) AND NATURAL SYSTEM MODEL (NSM) 37

APPENDIX C
    BIOGRAPHICAL SKETCHES OF MEMBERS OF THE COMMITTEE ON RESTORATION OF THE GREATER EVERGLADES ECOSYSTEM 39

Executive Summary

Purpose

This report evaluates scientific components of the Florida Bay studies and restoration activities that potentially affect the success of the overall Comprehensive Everglades Restoration Plan (CERP). Specifically, this report deals with scientific aspects of Florida Bay that feed back to the Everglades components of the CERP and are integral to the success of the overall restoration plan. It summarizes the science needed to determine potential long-term effects of Everglades restoration on the nature of the Bay. It is not intended to be a comprehensive review of the peer-reviewed literature on Florida Bay.

The Florida Bay and Adjacent Marine Systems Science Conference (Key Largo, April 2001) was a major stimulus for this report. Many of the CROGEE members attended the conference and participated in discussions with researchers working on Bay issues. This report is based on those discussions, reviews of poster presentations and abstracts from the conference, and an independent evaluation of pertinent peer-reviewed literature.

Description of the Bay

Florida Bay is a large, shallow marine ecosystem immediately south of the Everglades. It may be thought of as the marine extension of the Florida peninsula, beginning where the gently sloping land surface of the southern Everglades descends below sea level. It is bounded on the east and south by the Florida Keys and on the west by the Gulf of Mexico. It covers about 850 square miles (2200 square kilometers), mostly within Everglades National Park and the Florida Keys National Marine Sanctuary. It is dotted with several hundred small, mangrove-rimmed islands.

The average depth of Florida Bay is less than one meter and it is generally well mixed vertically. Shallow carbonate banks divide it into semi-isolated basins that restrict circulation, especially in the central and eastern zones. Its salinity is a function of direct rainfall and evaporation, inflow of Gulf of Mexico and Florida shelf water across its open western boundary, inflow of fresh water into the northeastern Bay by sheetflow and creeks fed by Taylor Slough and the C-111 canal (a major water conveyance canal, now intentionally blocked and the levees removed, on the southeast edge of Everglades National Park), outflow through the Keys to the southeast, and saline groundwater discharge. Cells of hypersaline water are common during the dry season.

For at least several decades until the late 1980s, clear water and dense seagrass (largely *Thalassia testudinum*, commonly known as turtle grass) meadows characterized most of Florida Bay. However, beginning around 1987, the turtle grass beds began dying in the central and western Bay, for reasons that remain uncertain and controversial. Water in the central and western Bay, which had been clear during recent decades, has become turbid because of phytoplankton blooms (which can also cause fish kills by consuming dissolved oxygen and releasing toxins) and sediment resuspension. A decline in fishing success for some species that use the Bay as a juvenile nursery habitat also was reported during that period.

Florida Bay is linked intimately to the Everglades. Some of the water draining from the Everglades through Taylor Slough/Craighead Basin flows directly into Florida Bay, supplying it with freshwater runoff. Additional water from the Everglades appears to reach the Bay indirectly after it is discharged from Shark River Slough to the northwest and mixes with inshore shelf water. The Everglades has been altered greatly during the past century, including the construction of canals, levees, pumps and control structures, and conversion of land to cities and farms. This has led to water quality degradation, nutrient enrichment, loss of wetlands, and landscape fragmentation in various parts of the Everglades.

Florida Bay is included in the CERP through the Florida Bay & Florida Keys Feasibility Study (FBFKFS), which is to be conducted to assess the current conditions of the Bay and to determine the modifications needed to restore it. Research in Florida Bay is carried out by many academic and governmental institutions. One of the most important of these is the Florida Bay and Adjacent Marine Systems Science Program, formed by state and federal agencies having regulatory and/or scientific interest in this region.

Potential Effects of the CERP on Florida Bay

An important assumption often made by scientists and managers associated with the CERP, and by the public, is that the increased flows of water deemed necessary to restore habitats in the Everglades also will contribute to the restoration and enhancement of Florida Bay. This is because increasing frequency, severity, and duration of hypersaline conditions in parts of the Bay, and a decrease in the spatial and temporal extent of estuarine conditions, are thought by some scientists to have been major factors leading to a dramatic die-off of turtle grass around 1987.

For a number of reasons, these assumptions may not be correct. First, the evidence linking the turtle grass die-off to hypersalinity is equivocal and there is little agreement within the Florida Bay research community that this was the causative factor of the die-off. Second, direct, fresh surface water flow into northern Florida Bay (i.e., via Taylor Slough and Craighead Basin, etc.) is predicted to be about the same in 2050 relative to the current condition whether the CERP is implemented or not. If this is correct, there will be little effect on salinities in central Florida Bay and no relief to any associated ecological problems that may exist. On the other hand, recent research suggests that some percentage of the proposed significant increase in fresh surface water flow through Shark River Slough will ultimately reach the central Bay by passing across the western boundary of the Bay after mixing with shelf water.

It also is not clear how, or if, the CERP will affect the magnitude of groundwater fluxes to Florida Bay. At present, the freshwater-saltwater interface in the surficial aquifer system is inland of the Bay. However, if the CERP raises overall water levels in the southern Everglades, this interface may be pushed southward over time toward Florida Bay in certain areas along the coast, and could result in fresh groundwater discharge directly to the Bay. Even if the direct discharge remains saline, changes in the quantity of groundwater inputs may be important to nutrient fluxes in some parts of the Bay.

In addition to the uncertainties concerning the amount of fresh surface and groundwater that may enter Florida Bay because of the CERP, it is possible that an increase in water would also bring an increase in nutrient inputs. If this is the case, the biological and ecological effects of such an increase in nutrient loadings are unclear. While there is a broad scientific consensus that the growth of phytoplankton of eastern Florida Bay and seagrasses throughout the bay is phosphorus-limited, there is less agreement about the relative importance of nitrogen and phosphorus in limiting the growth of phytoplankton and macroalgae in the central and western Bay.

Florida Bay phytoplankton blooms appear to develop where nitrogen-enriched water from the eastern Bay and from land drainage mixes with relatively phosphorus-rich water of the western/central Bay. Thus, the higher natural or anthropogenic loadings of nitrogen and, perhaps, phosphorus that may accompany increasing freshwater fluxes from Shark River Slough could potentially increase the frequency, intensity, and duration of phytoplankton blooms in regions of the Bay where these waters mix. Once generated, such blooms may be spread over larger areas within the Bay or be carried through the Keys to the coral reefs. Complicating the analysis of such issues is the lack of a circulation model for Florida Bay to link the hydrodynamic and ecological response of Florida Bay to changes in Everglades hydrology and to provide decision support and analysis tools to restoration planners. Such a model needs reliable information on water and nutrient fluxes, and a good linkage to models of the Everglades proper.

Findings and Recommendations

Because the CERP is an ambitious and comprehensive enterprise, with a long time horizon, *it is critical that the CERP be responsive to new information as it becomes available from the extensive ongoing research and monitoring programs throughout south Florida.* The assumption of a fresher and "healthier" Florida Bay because of the CERP is a case in point that should be reexamined.

Major considerations leading to this position are as follows. (1) Although it remains debatable how much new freshwater flow will enter the Bay proper because of the CERP, recent physical observations demonstrate that there is commonly a hydraulic connection, albeit with seawater mixing and a time lag, between Shark River Slough discharge and the interior of Florida Bay. (2) Some fraction of the DON (dissolved organic nitrogen) that would accompany increased freshwater flows from the Everglades will likely be available, either directly or indirectly, to support undesirable algal blooms within Florida Bay. (3) Enhanced blooms of phytoplankton and/or macroalgae may reduce seagrass cover and expose sediments to resuspension. Such resuspension will increase turbidity within the Bay and contribute to additional seagrass loss.

The scientific evidence suggesting that the CERP, as proposed, will affect the marine environment in ways that are not fully understood, and that may be perceived as undesirable, is sufficiently persuasive that the issue should be the subject of a focused technical review and evaluation. This review should be carried out as an early activity within the FBFKFS so that the conclusions can have an influence on early stages of Everglades restoration planning. The results of this analysis need to be evaluated by resource managers and planners of CERP so that appropriate consideration and management actions can be taken.

The importance of this issue has been recognized by some members of the Florida Bay and CERP research communities, and several projects are underway or will soon begin that will be helpful in addressing it. Research in the following areas is particularly important:

• Some components of the water budget for the Bay are poorly known, but understanding these fluxes is critical for evaluating the water quality in the Bay. In particular, estimates of groundwater discharge to the Bay differ by many orders of magnitude. Although this discharge is saline it may be a significant source of nutrients. Moreover, the higher water levels produced by the CERP may affect the magnitude of these fluxes and, in extreme cases, may result in direct discharge of fresh groundwater to the Bay. Also of importance to both salinity and nutrient fluxes is a full characterization and quantification of surface runoff in the Taylor Slough, Craighead, and Shark River Slough basins, including the seasonality of flow. Diffuse seepage through the Buttonwood Embankment should also be investigated.

• On the basis of modeling, annual freshwater flows through the Taylor Slough/Craighead Basin region to Florida Bay will be about the same in year 2050 under the expanded CERP scenario D13R4 compared to the simulated current condition. At the same time, annual freshwater flow through Shark River Slough, some of which migrates to the Bay indirectly and with a time lag, is projected to increase by almost 80%. The effects on nitrogen and phosphorus fluxes of increasing this freshwater need to be quantified.

- Total loads of nitrogen and phosphorus from freshwater sources should be estimated as accurately as possible. A better understanding of the transport, bioavailability, and rates of transformation of DON and DOP (dissolved organic phosphorus) into forms that can be used by algae and macroscopic aquatic plants is needed to provide insight into the effects of increases in nutrient-bearing freshwater flows to the Bay. Quantifying the magnitude of nutrient loadings by source (e.g., organic soils oxidation, urban and agricultural runoff, and regional atmospheric deposition) also will become relevant if steps to reduce nutrient loading to the Bay become necessary.

- A historical characterization of the Bay's water quality would be very useful for a perspective on restoration goals. Such a characterization would be based on anecdotal as well as any scientific information available.

- Currently there is no Florida Bay circulation model suitable for research and management purposes, although there are several candidates; such a model is essential to support a Bay water quality model and thus facilitate analysis of CERP effects on the Bay. The difficult and time-consuming tasks of selection, development, and application of a circulation model and water-quality model for the Bay should be key components of the FBFKFS.

- To evaluate the effects of CERP on Florida Bay, there must be a linkage of the output of the South Florida Water Management Model (SFWMM), which has a southern boundary of the mangrove zone, and input to the Bay models. One possibility for bridging this gap is the U.S. Geological Survey's (USGS) Tides and Inflows in the Mangroves of the Everglades (TIME) model; another is the similarly structured South Florida Water Management District (SFWMD) South Florida Regional Simulation Model (SFRSM). Both models are still under development. The USGS, SFWMD, and Corps of Engineers have not reached consensus on how to effect the interfacing of hydrologic modeling and circulation modeling, and both the USGS and SFWMD modeling efforts are proceeding without such an agreement. An interagency agreement on which model will most usefully serve as an interface between landside hydrologic and Bay hydrodynamic modeling is needed.

- Estimates of the influence of the CERP on Florida Bay should also be inferred from statistical and time series analysis of existing data and/or use of simpler "box models," such as FATHOM (Flux Accounting Tidal Hydrology Ocean Model). This is especially important given the technical difficulties involved in developing full-scale simulation models.

- Human factors such as population growth and economic activity; and environmental events whose drivers are distant from Florida Bay and unrelated to CERP activities, such as hurricanes, flooding of the Mississippi River, atmospheric deposition of nutrients from emission sources remote to Florida Bay or its watershed, and sea-level rise; also may influence local conditions in Florida Bay. Research is needed to better define these potential effects and to integrate the results into predictive ecosystem-response models.

There is much to commend in the overall research effort on Florida Bay and in the goals of the FBFKFS. But the evidence that the CERP will cause changes in Florida Bay that may be perceived as undesirable is sufficiently persuasive that the process of more detailed evaluation should begin as an early part of the FBFKFS. The effort required for all these tasks is daunting. Sufficient time and resources should be made available as part of the FBFKFS for essential research to ensure its success and usefulness for management decisions.

1

Introduction

Florida Bay is a large, shallow marine ecosystem immediately south of the Everglades (Figure 1). It may be thought of as the marine extension of the Florida peninsula, beginning where the gently sloping land surface of the southern Everglades descends below sea level. It is bounded on the east and south by the Florida Keys and on the west by the Gulf of Mexico, and covers about 850 square miles (2200 square kilometers) including about 700 square miles (1800 square kilometers) within Everglades National Park. Much of the rest is in the Florida Keys National Marine Sanctuary. It is dotted with several hundred small islands, many of which are rimmed with mangroves.

The Physical System

The average depth of Florida Bay is less than one meter (CERP, 2001) and it is generally well mixed vertically. Shallow carbonate banks subdivide it into semi-isolated basins that restrict circulation, especially in the central and eastern zones (Fourqurean and Robblee, 1999). Its salinity is highly dependent on local rainfall and evaporation; cells of hypersaline water are common during the dry season. At the discharge locations of the Taylor Slough and C-111 basins (Figure 2) in the northeastern Bay, the salinity is much lower during the wet season than in the dry season.

The northern boundary, which represents the primary interface between the Bay and the up-gradient ecosystems of the Everglades, is complex (Figure 2). Water carrying dissolved and particulate matter flows in shallow sloughs that become channelized through broad bands of creeks as they near the coastline (Rudnick et al., 1999). Fresh water enters the Bay primarily as channel flow fed by Taylor Slough and to a lesser extent the C-111 canal (a major water conveyance canal, now intentionally breached, on the southeast edge of Everglades National Park) to approximately 20 microtidal creeks that cut through the Buttonwood Embankment and as sheetflow across marl prairies of the southern Everglades (Figure 1) (Davis et al. 2002; CERP, 2001). Groundwater discharge along this boundary is poorly quantified but is primarily saline (Fitterman and Deszcz-Pan, 2001). The quantity and distribution of material exchanges between the Everglades watershed and the Bay are not well understood (Sklar and Browder, 1998).

The western boundary of the Bay is open to the Gulf of Mexico (Fourqurean and Robblee, 1999). A west-to-east (or NW-to-SE) current generated by winds, tides, and the broader circulation of the Gulf of Mexico, Caribbean Sea, and Atlantic Ocean generally prevails across the Gulf-Bay boundary (Lee et al., 2002; Smith and Pitts, 2002). It is, therefore, very probable that some of the water and nutrients discharged from Shark River Slough (Figures 1 and 2) migrate south around Cape Sable, across this boundary, and into the Bay (Boyer et al., 1999). Net outflow to the Atlantic Ocean occurs along the southern boundary through passes between the Florida Keys (Smith and Pitts, 2002). The volumes are lower now than they were before construction of the Flagler Railroad, which extended from the mainland to Key West, and was completed in 1912.

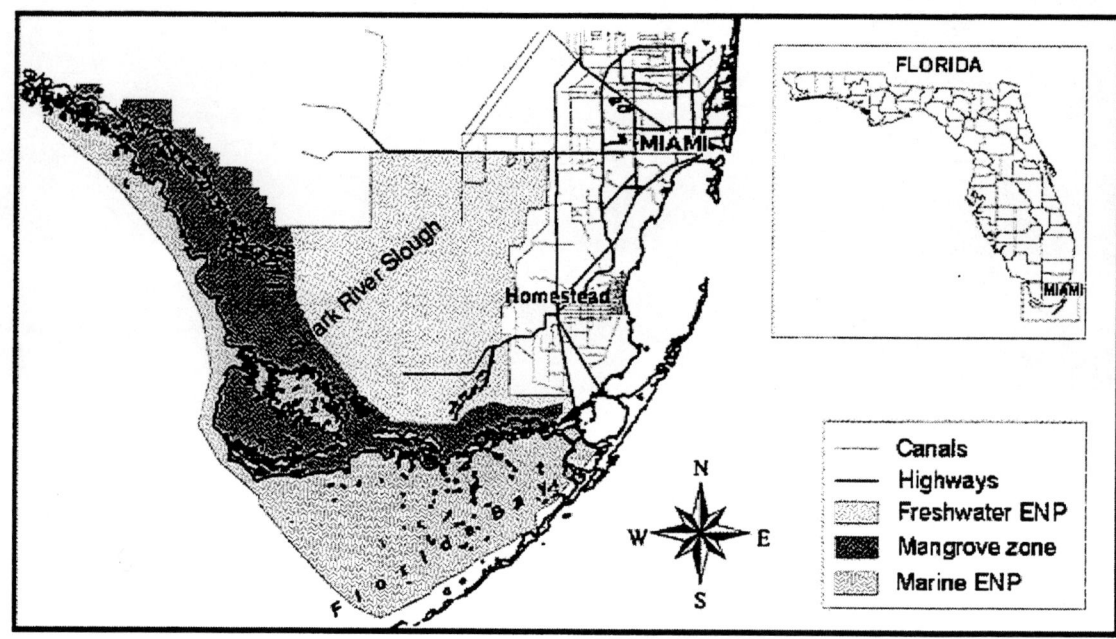

FIGURE 1 Location of Florida Bay with respect to other major features of South Florida. (Note: ENP is Everglades National Park.) Source: Florida Coastal Everglades Long Term Ecological Research, 2002.

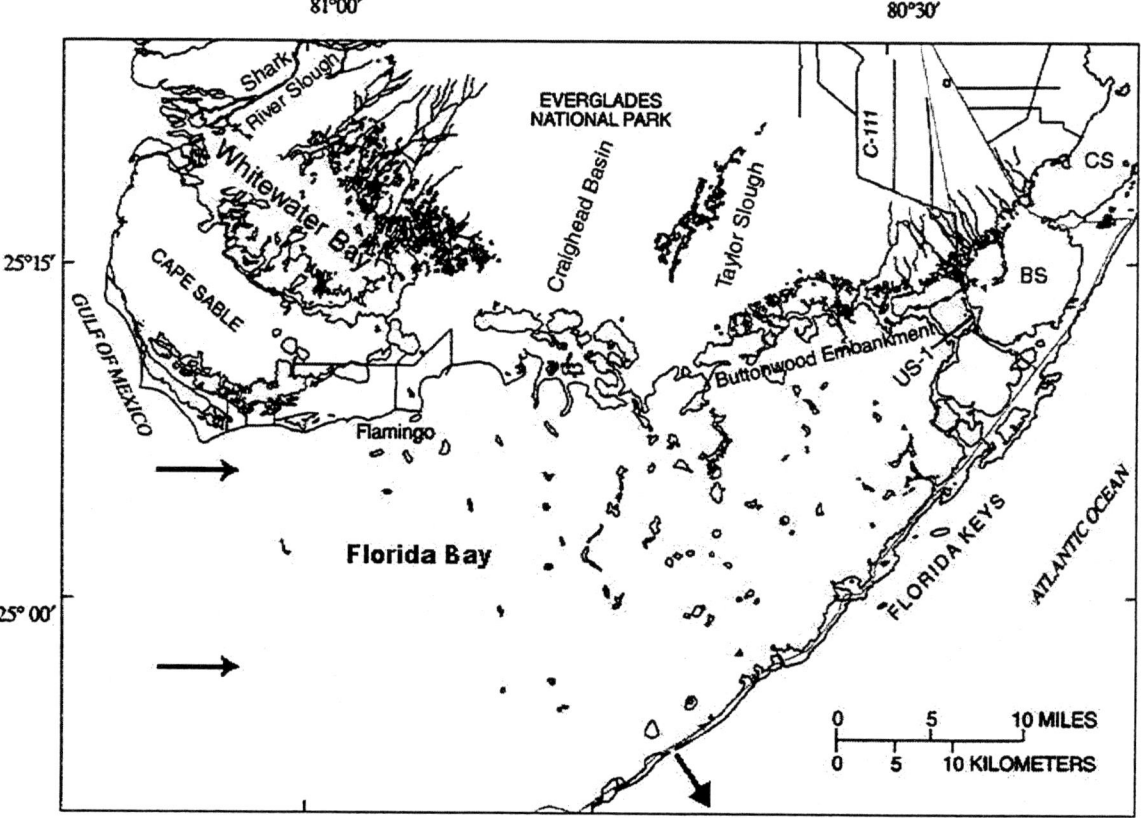

FIGURE 2 Detail map of Florida Bay. Arrows indicate generalized circulation pattern through Florida Bay (after Smith and Pitts, 2002). (Note: BS is Barnes Sound; CS is Card Sound. Biscayne Bay is to the northeast of Card Sound, off the map). Source: Hittle, 2001.

Nutrients in the Bay

In terms of nutrient status and water chemistry, Florida Bay can be divided into three zones (Boyer et al., 1997). These are (a) a western zone of fairly stable marine conditions and nitrogen:phosphorus ratios in seston (i.e., in particulate matter in the water column) approximating the Redfield ratio (indicating balanced conditions relative to the nitrogen and phosphorus needs of the phytoplankton community); (b) a central zone that is often hypersaline, has high nitrogen:phosphorus ratios for seston, and high levels of dissolved organic matter; and (c) an eastern zone with variable salinity, high concentrations of dissolved inorganic nitrogen, and even higher nitrogen:phosphorus ratios for seston. These high nitrogen:phosphorus ratios suggest that phytoplankton growth in at least the eastern Bay is limited by phosphorus.

For at least several decades until the late 1980s, most of Florida Bay was characterized by clear water and dense seagrass (largely *Thalassia testudinum*, or turtle grass) meadows (CERP, 2001), albeit with some interannual variation (Halley, 2002). However, beginning around 1987, the seagrass beds began dying in central and western Florida Bay (Fourqurean and Robblee, 1999), for reasons that remain a subject of scientific controversy. Several ecological changes that may be related to the die-off have ensued in the Bay. Water in central and western Florida Bay, which had been "gin clear" during recent decades, has become turbid, in part because of phytoplankton blooms and in part because of increased sediment resuspension (Zieman et al., 1999). A decline in fishing success for some commercial and recreational species that use the Bay as a juvenile nursery habitat also was reported in that period (CERP, 2001).

Linkages with the Everglades

Florida Bay is intimately linked to the Everglades from both hydrologic and public policy perspectives. From a hydrologic perspective, some of the water draining from the Everglades flows directly into Florida Bay, supplying it with freshwater runoff that is high in total nitrogen (Brand, 2002) and low in total phosphorus (Rudnick et al., 1999). Other freshwater drainage appears to reach the Bay indirectly after it is discharged from Shark River Slough and mixes with southwest Florida shelf water. The Everglades has been greatly altered during the past century by human activity, including the construction of canals, levees, pumps and control structures, and conversion of land to agricultural and urban uses. This has led to vast ecological and hydrologic changes within the Everglades, including degradation of water quality, nutrient enrichment, landscape fragmentation, invasion by exotic species, declines in native plant and animal species (McPherson and Halley, 1996), and an overall loss of about half of the original wetland area (Davis et al., 1994). Because of the intimate hydrologic connection between the Everglades and the Bay, restoration of the Everglades most likely will affect conditions in the Bay.

From a public policy perspective, Florida Bay is connected with the rest of the Everglades system because it is a part of Everglades National Park and is included in the Comprehensive Everglades Restoration Plan (CERP). CERP planners acknowledge that many questions about the Bay and its relationship to the Everglades have no definitive answers at present. The CERP therefore recommended that the Florida Bay & Florida Keys Feasibility Study (FBFKFS) be conducted to assess the current conditions of the Bay and to determine the modifications needed to restore it (CERP, 2001). The FBFKFS is being conducted under the authority of the Water Resources Development Act (WRDA) of 1996, which allows for continuation of studies and analyses that are necessary to further the CERP (CERP, 2001).

Current Research and Oversight

Research in Florida Bay is carried out under the aegis of many academic and governmental institutions. The Florida Bay and Adjacent Marine Systems Science Program (http://www.aoml.noaa.gov/flbay/) is a collaborative effort of federal, state, and local agencies that conduct and sponsor closely complementary research, monitoring, and modeling projects on the Bay ecosystem. The program also provides a valuable mechanism for meetings of research teams, and it sponsors topical workshops and the important Florida Bay and Adjacent Marine Systems Science Conferences, the most recent of which (April 2001) provided the impetus for this report.

A Program Management Committee (PMC), composed of representatives of 11 state and federal agencies, guides the program. The PMC also provides for independent expert review through the Florida Bay Science Oversight Panel, as defined in the 1994 interagency science plan (Armentano et al., 1994). That plan defined the Panel's role to include both technical and management review of program development, quality of research, modeling, and monitoring. Ad hoc advisory panels of experts in specialized subject areas also are formed from time to time. Reports of the Science Oversight Panel, along with many other reports on Florida Bay, may be found at http://www.aoml.noaa.gov/ocd/sferpm/oid.html.

Many academic institutions are involved in Bay research, most of which is funded by the PMC agencies. Some now also collaborate through the Florida Coastal Everglades Long Term Ecological Research (FCE LTER) program, which began in 2000. Research associated with the FCE LTER focuses on the coastal Everglades to investigate how nutrient availability and cycling interacts with the hydrologic regime to potentially control ecosystem structure and function (http://fcelter.fiu.edu/).

Report Objective and Overview

This report evaluates scientific components of the Florida Bay studies and restoration activities that potentially affect the success of the overall CERP. It deals with scientific aspects of Florida Bay that feed back to the Everglades components of the CERP and are integral to the success of the overall plan. In particular, it focuses on the potential long-term effects of Everglades restoration on the nature of the Bay. The report reflects discussions and reviews of poster presentations and abstracts from the April 2001 Florida Bay Science Conference and a review of relevant peer-reviewed literature.

Chapter 2 of this report describes the potential effects of the CERP on Florida Bay, specifically addressing the question: will increased flows of water deemed necessary to restore terrestrial and wetland habitats in the Everglades contribute to the restoration and enhancement of Florida Bay or will they have detrimental effects on the Bay ecosystem? Chapter 3 elaborates on the kinds of research that will be required to shed light on this question. Chapter 4 summarizes the committee's conclusions and recommendations.

2

Potential Effect of the Comprehensive Everglades Restoration Plan (CERP) on Florida Bay

An important assumption often made by scientists and managers associated with the Comprehensive Everglades Restoration Plan (CERP), and by the public, is that the increased flows of water deemed necessary to restore Everglades habitats also will contribute to the restoration and enhancement of Florida Bay (e.g., SFWMD, 1998). This assumption appears to rest on three pillars: (1) that there has been an undesirable (but poorly quantified) long-term increase in salinity in the Bay as a result of water management practices beginning around the 1920s (McIvor et al., 1994) and possibly also as a result of changes in circulation because of railroad construction in the Keys during the early 1900s (Halley and Roulier, 1999); (2) that the increase in salinity resulted in increasing frequency, severity, and duration of hypersaline conditions in parts of the Bay, and a corresponding decrease in the spatial and temporal extent of oligohaline-mesohaline conditions; and (3) that the restoration plan will ameliorate this problem by delivering more freshwater to the Bay than it currently receives. Hypersalinity is undesirable, at least in part, because it is thought by some researchers to have been a major factor leading to a dramatic die-off of *Thalassia* around 1987.

While the *Thalassia* die-off was first attributed to a direct salinity effect (Robblee et al., 1991), various alternative hypotheses have been advanced more recently. These include hypoxia and sulfide toxicity (Carlson et al., 1994), loss of the estuarine nature of the Bay, overdevelopment of seagrass beds, abnormally warm late summer and fall water temperatures, sedimentation due to a lack of severe storms (Zieman et al., 1988), eutrophication (Lapointe and Clark, 1992), pathogens (Durako and Kuss, 1994), and the decline of herbivory due to the decline of green turtle and manatee populations (Jackson et al., 2001). Of these hypotheses, hypersalinity and eutrophication are mostly likely to be influenced by the CERP.

The view that increasing freshwater flow to the Everglades and to Florida Bay is a "win-win situation" may not be entirely correct. Brand (2002) concluded that "if more freshwater from the Everglades agricultural system is pumped into Florida Bay, as proposed, the algal blooms will increase and the ecological problems of Florida Bay will get worse, not better." Brand's concern was not with the freshwater itself, but with the nutrients, especially nitrogen, that it would contain. Another review of recent changes in the ecology of Florida Bay and the coral reef communities of the Florida Keys also concluded that excess nitrogen inputs were responsible for many undesirable changes in the marine system, and that the CERP had not addressed this issue adequately (Lapointe et al., 2002). Furthermore, Prager and Halley (1999) and Halley (2002) cautioned that estuarine conditions, which might result from restoring freshwater flows to the Bay to their historic levels could lead to increased turbidity from suspended sediment and loss of the "gin clear" bay remembered to have occurred in the 1960s and 1970s. Indeed, anecdotal evidence exists for generally turbid conditions in Florida Bay 100 years ago. Gregg (1902) stated that "Florida Bay…is a shallow bay, and the water is usually roily and whitish from the disintegration of the coral rock," and "Now we will have quite a sail through milky water, until we near

Tavernier Creek." Thus, Halley (2002) described the potential increase in turbidity as "simply a natural response to creating a bay more like the one that existed at the turn of the century." This vision, however, is different from the view that associates pristine conditions in the Bay with clear water.

This chapter first examines anticipated changes in patterns of freshwater inflows to the Bay under the CERP, and then considers whether these changes, which are designed to lessen the likelihood of hypersaline conditions developing in the Bay, may be undermined by the accompanying influx of nutrients.

Freshwater Inflows to the Bay

Simulating past, present, and future freshwater inflows to Florida Bay is challenging due to the large number and tidal nature of the creeks that drain the coastal mangrove zone, and to poor linkages between Everglades and Florida Bay models (see Chapter 3). Some flow measurements are now being made (e.g., Hittle et al., 2001), but flows are more commonly modeled. To date, the Natural System Model (NSM) has been the primary tool for simulating the hydrologic behavior of the pre-drainage Everglades. The South Florida Water Management Model (SFWMM) has been used to simulate the system infrastructure and operations as they are currently ("1995 Base"), as they will be in 2050 without any CERP projects in place ("2050 Base"), and as they could be in 2050 with CERP projects completed ("D13R"). The models are discussed in Chapter 3, and the model runs are described in more detail in Appendix B.

Taylor Slough and Craighead Basin

Direct fresh surface water flow into Florida Bay occurs primarily through Taylor Slough and, to a lesser extent, Craighead Basin. These modeled flows are shown in Figure 3. Slightly to the east, southward flow through the C-111 canal to the Eastern Panhandle area of Everglades National Park (Figure 3, right-most set of bar graphs) also occurs. Most of this flow, however, does not end up in Florida Bay. Rather, it is discharged east of U.S. Hwy 1, eventually passing into Barnes Sound and thence northeastward through Card Sound to Biscayne Bay. Very minor quantities of Eastern Panhandle water discharge to northeastern Florida Bay through the degraded C-111 embankment (Richard Punnett, USACE, Personal commun., July 2002).

From Figure 3 it can be seen that more water flows into the Eastern Panhandle under the 1995 Base or "current condition" relative to the NSM simulation. This is reduced under alternative D13R4 (a scenario based on D13R that would capture additional water "lost" to tide). The surplus (119,000 - 50,000 = 69,000 acre-ft per year (8.5 x 10^7 cubic meters per year)) would be redirected upstream into Shark River Slough and downstream to a very small degree into Craighead Basin and Taylor Slough, in an effort to replicate NSM flows (although CERP flow targets may differ from NSM estimates). Hence, simulated fresh surface water flows to Florida Bay through Craighead Basin and Taylor Slough increase only minutely from 32,000 + 94,000 = 126,000 acre-ft per year (15.5 x 10^7 cubic meters per year) under the 1995 Base scenario to 45,000 + 82,000 = 127,000 acre-ft per year (15.6 cubic meters per year) under the D13R4 scenario. The Taylor Slough flow under the D13R4 scenario would be close to the NSM-estimated flow shown in Figure 3.

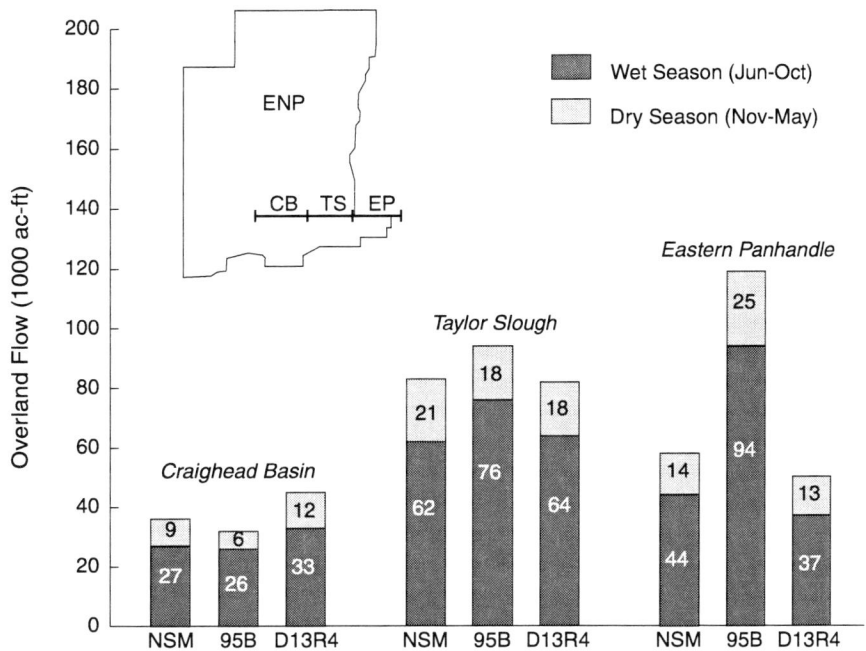

FIGURE 3 Average annual overland flows toward Florida Bay across Craighead Basin, Taylor Slough and Eastern Panhandle for the 31-year simulation. Comparison of surface water flows in eastern Everglades National Park (ENP). Flow is generally southward across the Craighead Basin-Taylor Slough-Eastern Panhandle (CB-TS-EP) cross-section (shown in inset map of northeastern ENP; see Figure 1 for larger map). CB, TS, and EP are each an eight-mile long segment of the cross-section. Output is shown for three different simulations: Natural System Model (NSM), 1995 base or "current condition" (95B) and D13R4 (a variation of D13R, which is the year 2050 CERP simulation from USACE, 1999). Short descriptions of these are given in the text; more complete definitions are given in Appendix B. Note: Note: NSM water depths at key ENP gage locations are used as operational targets for most alternatives. NSM flows are NOT targets and are shown for comparative purposes only. Source: USACE, 2002.

The total annual flow is, of course, only one aspect of the data. The distribution of flow between the wet and dry seasons (shown in the Figure 3 graphs) may be important as well. In addition, the simulated *averages* do not reflect annual variability in discharges; this must be accounted for when analyzing the full 31-year output from the SFWMM and its ultimate interface with Florida Bay modeling. Finally, changes to rainfall-based water management practices that occurred between the mid-1980s and 1995 resulted in increasing the amount of freshwater flow into eastern Florida Bay, relative to rainfall, at least since 1993 (Sklar et al., 2002). The 1995 base is a simulation for the present period since those changes occurred; it does not represent conditions of the preceding decades, which are less well known.

Overall, however, total fresh surface-water inflows to Florida Bay via Craighead Basin and Taylor Slough are predicted to be about the same with CERP as under current conditions. If these predictions are correct, the salinity of this region of the Bay may not change materially. The lack of an operational hydrodynamic model (Chapter 3) increases the uncertainty of such predictions.

Shark River Slough

In contrast to these minor proposed changes to flows in Craighead Basin and Taylor Slough, the CERP plans a dramatic increase in flow down Shark River Slough relative to the current condition (Figure 4). This is significant, because recent measurements strongly suggest that some of this flow eventually will reach the inner parts of Florida Bay. A detailed review by Smith and Pitts (2002) indicated, "a generally west-to-east movement of water through the interior of the Bay that eventually exits through the tidal channels between keys on the southeastern and southern sides of the bay." These authors concluded their analysis of 15 years of physical observations by emphasizing that "[a]veraging over tidal periods and the longer time scales associated with meteorological forcing...reveals transport pathways that represent a clear coupling between Gulf and Atlantic sides of the Keys. Gulf-to-Atlantic transport can be either around the Keys...or it can involve a more complex route through Florida Bay and the tidal channels..." Numerous drifter studies have shown that Shark River Slough water tends to pass along the western boundary of Florida Bay and must often have access to the central Bay (Lee et al., 2002). Boyer et al. (1999) believed that they could see the effect of "a freshening of the waters of the southwest Florida Shelf from Shark Slough drainage" on salinity declines in western Florida Bay. Additional qualitative information about the linkage of the Bay with Shark River Slough discharge is provided by D'Sa et al. (2002) on the basis of remote sensing of salinity patterns. These researchers concluded that "Gulf waters entering the bay primarily from the northwest (near East Cape) entrain freshwater from the Shark River Slough and other smaller rivers in southwest Florida as indicated by the lower salinities observed in the vicinity of Cape Sable..." (Figure 2).

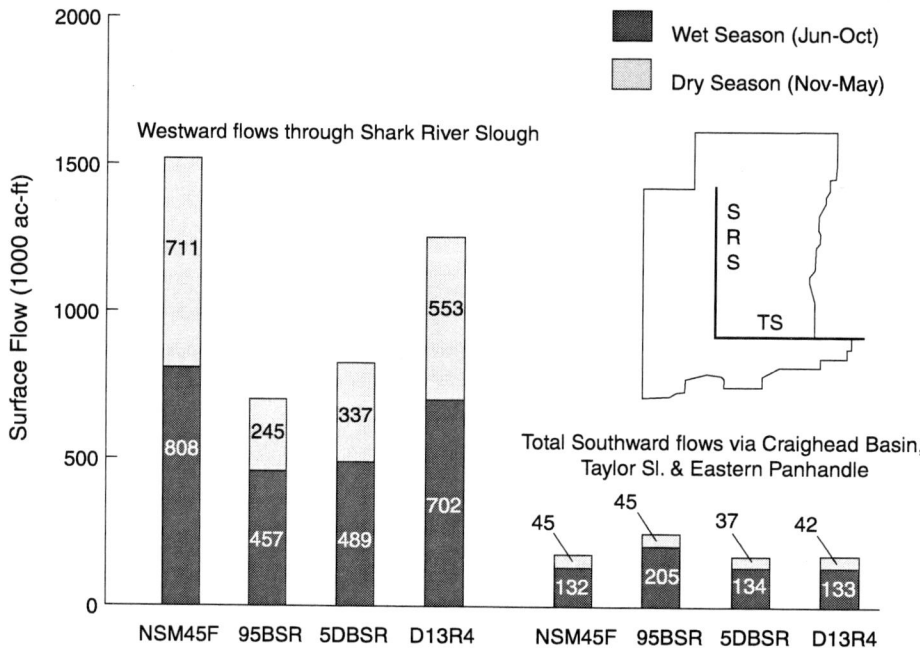

FIGURE 4 Average annual overland flows toward Whitewater Bay and Florida Bay for the 31-year simulation period. Comparison of flows across Shark River Slough (SRS) and Craighead Basin/Taylor Slough/Eastern Panhandle (TS) cross-sections for different modeled scenarios. (Cross-section locations are shown in inset map of northeastern Everglades National Park; see Figure 1 for larger map.) Definitions of NSM45F (Natural System Model), 95BSR (1995 base or "current condition"), 50BSR (2050 base or "without project condition"), and D13R4 are given in Appendix B. The TS data include Eastern Panhandle flows that discharge to water bodies other than Florida Bay (see Figure 3 for details); total flows to Florida Bay for the 1995 base are actually similar to those of the other model runs. Note: NSM water depths at key ENP gage locations are used as operational targets for most alternatives. NSM flows are NOT targets and are shown for comparative purposes only. Source: USACE, 1999. (Also available at http://www.everglades-plan.org/pub/restudy_eis.shtml.)

Groundwater

The long-term effect of the CERP on the magnitude and salinity of groundwater fluxes to Florida Bay is uncertain. At present, the freshwater-saltwater interface in the surficial aquifer system is at least six kilometers inland of the Bay (Fitterman and Deszcz-Pan, 2001). The salinity of shallow groundwater along the coast of the mainland is close to that of seawater (Fitterman and Deszcz-Pan, 2001). For this reason, discharge of fresh groundwater from the Everglades actually occurs on the mainland, and subsequently enters the Bay as surface runoff. The discharge volume is estimated to be small relative to that of surface water by CERP modelers (USACE, 1999). However, if the CERP raises overall water levels in the southern Everglades, the freshwater-saltwater contact in the surficial aquifer system may be pushed southward over time toward Florida Bay in certain areas along the coast, and could result in fresh groundwater discharge directly to the Bay. This might result in a small net increase in freshwater from this source because the loss due to evapotranspiration during the transit between inland groundwater discharge and the Bay would be eliminated.

Although discharge of fresh groundwater to the Bay appears to be negligible under current conditions, there have been several attempts to estimate the magnitude of saline groundwater discharge. Saline discharge can be the result of buoyant counterflow of saline porewater induced by flow in a freshwater lens (Kaufman, 1994), geothermal convection induced by temperature differences between ocean water and fluids in the interior of a carbonate platform (Sanford et al., 1998), or local recirculation driven by waves and tidal pumping (Li et al., 1999). Recent estimates are based on the chemical tracers ^{222}Rn, CH_4, and ^{4}He, and seepage meters (Corbett et al., 1999, 2000; Top et al., 2001). Corbett et al. (1999, 2000) estimated a discharge of 1 to 3 centimeters per day, which implies a flux for the Bay as a whole of 7 to 22 x 10^9 cubic meters per year. Top et al. (2001) reported even higher estimates of 6 to 12 centimeters per day, or about 45 to 85 x 10^9 cubic meters per year. These values are many times larger than the direct surface freshwater flow into northeast Florida Bay in the very wet year of 1997-1998 (0.24 x 10^9 cubic meters per year; Patino and Hittle, 2000). It is difficult to imagine how this modest freshwater input could have resulted in the observed reductions in salinity in north-central and eastern Florida Bay if there were a large net input of saline groundwater. The upper estimates of saline groundwater flow would displace a one meter deep water column in 8 to 16 days, while the average residence time of water in the isolated basins of this region of the Bay is likely on the order of months (George Jackson, Texas A&M University and Ned Smith, Harbor Branch Oceanographic Institution, personal commun., April 2002).

All of these estimates seem unreasonably high unless these fluxes are dominated by recirculation of Bay water through shallow aquifers under and adjacent to the bay. A comparative study of groundwater flux estimation techniques in the northeastern Gulf of Mexico (Burnett et al., 2002) noted that estimates from seepage meters and tracer measurements may include flow due to tidal pumping and wave action, which are generally not included in estimates from steady-state groundwater flow and transport models. Such recirculation might be an important mechanism to transfer nutrients from the aquifer to the Bay, but it would not represent a net input of water. Nevertheless, discharge of recirculated saline groundwater to the central and eastern Bay may influence local water budgets, especially during droughts.

Nutrient Fluxes to the Bay

In the previous section, it was shown that there are significant unknowns with respect to the volumes and spatial patterns of freshwater flows to Florida Bay that will result from the restoration effort. In this section, the possible effects on the Bay of the nitrogen and phosphorus that may accompany those flows is discussed.

There appears to be consensus that the Everglades wetlands intercept large quantities of phosphorus and that the phytoplankton of eastern Florida Bay are phosphorus limited. Seagrasses throughout the Bay appear to be phosphorus limited (Fourqurean et al., 1992). There is less agreement about nutrient issues in the central and western Bay. There, the ratio of dissolved inorganic nitrogen (DIN) to dissolved reactive phosphate (DIP) usually suggests strong phosphorus limitation of phytoplankton growth (Boyer et al. 1999), but many bioassay measurements indicate that nitrogen is the nutrient of chief concern for phytoplankton and macroalgal blooms (Brand, 2002; Tomas et al., 1999). It is difficult to reconcile these two lines of evidence unless a significant fraction of the dissolved organic phosphorus (DOP) in Florida Bay is also accessible to the phytoplankton while much of the DON is not. Brand (2002) has developed this argument and summarized data indicating that phytoplankton blooms develop in the zone where relatively nitrogen-rich but phosphorus-poor water from the eastern Bay mixes with the phosphorus-rich but nitrogen-poor water flowing south along the inner southwest Florida shelf.

There also is strong evidence that nitrogen (mostly in dissolved organic form) and phosphorus are exported as a function of water flow from the Everglades (Rudnick et al., 1999). Thus, it seems likely that increasing water flows through Shark River Slough will result in a larger net flux of nitrogen and phosphorus out of the Everglades. It is interesting with regard to the CERP that Rudnick et al. (1999) also found an apparent net increase in phosphorus concentration as water from Shark River Slough passed through the mangrove zone. While the source of this phosphorus remains unknown, it persisted through the three-year period they analyzed. Brand (2002) hypothesized that the relatively elevated phosphorus concentrations of southwest Florida shelf water may be from phosphorite-rich quartz sand deposits that underlie much of the northwestern Bay. The hydraulic potential in the underlying Upper Floridan aquifer is higher than that of the Bay (Bush and Johnston, 1988), which would force groundwater upward through these sand deposits.

A major uncertainty with regard to effects of CERP-modified water deliveries to the Bay on nutrient loading/limitation in the Bay concerns the role of dissolved organic nitrogen (DON) in promoting the growth of phytoplankton and macroalgae. DON constitutes by far the largest fraction of the total nitrogen concentration in both the freshwater sources to the Bay and in the Bay water itself (Rudnick et al., 1999, Boyer et al., 1997).

Until recently, research on the role of nitrogen in Florida Bay focused primarily on dissolved inorganic nitrogen (DIN). However, globally 70 percent of the dissolved nitrogen transported to the sea in rivers is DON (Maybeck, 1982). In temperate regions, the proportion of the total dissolved nitrogen (TDN) pool consisting of DON rises during warm summer months when nitrate concentrations decrease in freshwater (Pardo et al., 1995) and marine (Carlsson and Graneli, 1998) environments. This suggests that in subtropical regions like south Florida, DON may be an even greater fraction of the TDN.

Worldwide, few studies have evaluated the bioavailability of DON, and even fewer have characterized DON in terms of its chemical structure (Stepanauskas et al., 1999). DON is commonly defined operationally as the organic nitrogen that passes through a filter with a nominal pore size of $0.45\ \mu$. Thus defined, DON includes the colloidal fraction (including some bacterial biomass that is rich in nitrogen), as well as molecules that are in a truly dissolved state. Although most of the nitrogen in organisms occurs in amino acids, only a small fraction of DON, perhaps 5-10 percent, can be identified as rapidly cycling free amino acids, amines, and urea (McCarthy et al., 1997). This fraction (i.e., amino acids and urea) may be used by some algae (Antia et al., 1991); higher molecular weight DON must be mineralized for the nitrogen to become available to phytoplankton, macroalgae, or submerged aquatic plants. The majority of DON appears to be "contained in amide functional groups, suggesting that most long-lived DON is hydrolysis-resistant or composed of recalcitrant non-protein amide-containing biochemicals" (McCarthy et al., 1997) raising questions about the bioavailability of high molecular weight DON.

The prevailing view that DON exported from terrestrial systems is largely refractory has been challenged by recent studies. Several investigators have shown that heterotrophic bacterioplankton mineralize humic-bound DON (Carlsson and Granéli, 1998; Carlsson et al., 1993; Bushaw et al., 1996; Carlsson et al., 1999), and Seitzinger and Sanders (1997) demonstrated that up to 80% of total DON was metabolized by bacteria in Hudson and Delaware River water. Furthermore, in well-mixed coastal systems like Florida Bay, photochemical breakdown of DON has been shown to enhance DON availability to microbes (Bushaw-Newton and Moran, 1999; Tarr et al., 2001; Wiegner and Seitzinger,

2001; Koopmans and Bronk, 2002). Recent laboratory work by Seitzinger et al. (2002) using water from Barnegat Bay, a New Jersey estuary, showed complex, non-linear relationships between phytoplankton production and DON addition. DON bioavailability varied seasonally, but overall, urban/suburban storm water runoff had a higher proportion of bioavailable DON (59% ± 11) than pastures (30% ± 14) and forests (23% ± 19). Preliminary chemical analyses of Everglades DON by Rudolf Jaffe at Florida International University showed that it contains a significant (5-10%) fraction of proteins and other compounds that should be readily available to bacteria (Rudolf Jaffe, FIU, personal commun., 2001). The relatively long residence time of water within eastern Florida Bay (on the order of months; see earlier references) in contrast to typical bacterial generation times as high as once per day (Joseph Boyer, FIU, personal commun., April 2002) suggests that residence times of this magnitude provide ample time for bacteria to mineralize a significant fraction of the DON entering this portion of the system.

Like DON, little is known about the reactivity of dissolved organic phosphorus (DOP) in coastal systems, although Brand (2002) has suggested that it is likely to be more available than DON. As with nitrogen, DOP makes up the greatest fraction of the total phosphorus entering Florida Bay in creek discharge from the Everglades (Rudnick et al., 1999). However, the contribution of this source of phosphorus to the Bay is likely only a small fraction of the Bay's phosphorus budget (Rudnick et. al., 1999). In addition to the large source of phosphorus entering Florida Bay from the Gulf of Mexico, internal Bay phosphorus-cycling (e.g., algal and bacterial alkaline phosphatase activity) (Beardall et al., 2001; Wright and Reddy, 2001) and macrophyte mining of sediment phosphorus (Jensen, et. al., 1998) could be significant.

A preliminary evaluation by Rudnick et al. (1999) of nitrogen and phosphorus inputs to Florida Bay from various sources deserves comment. Their analysis led the authors to conclude that the flux of nitrogen and phosphorus from the Everglades is so small compared with other sources that it likely would not affect the Bay if it increased modestly. Although the inventories of nutrient inputs that Rudnick et al. (1999) developed for the entire Florida Bay and its eastern portion provide a useful perspective for some issues, it is not clear that they are adequate to evaluate the finer-scale spatial and temporal patterns of the phytoplankton blooms. For example, while nutrient discharges from a sewage treatment plant may be a small part of the total nitrogen or phosphorus input to an estuary, they can have large impacts on the areas where the effluents are discharged.

Similarly, the higher loadings of nitrogen and phosphorus that may accompany increasing freshwater fluxes from the Everglades could increase the frequency, intensity, and duration of phytoplankton blooms in certain regions of Florida Bay, even though they would be small relative to nutrient flows through the Bay from the western border or from the atmosphere. Boyer et al. (1999) noted that increased freshwater flows from the Everglades to the eastern Bay could account for some water-quality changes, such as lower salinity in that region. Correlations between water discharge from the Everglades and phytoplankton blooms (chlorophyll *a*) in the north central Bay (Brand, 2002) also provide some circumstantial evidence that nitrogen enhancement of phytoplankton growth may be a consequence of increased freshwater flow from the Everglades. Once generated, such blooms may be spread over larger areas within the Bay or carried through the Keys to the coral reefs (Smith and Potts, 2002).

There is some reason to be concerned that impacts to Florida Bay that would be caused by any increased nitrogen or phosphorus discharges that may occur through Shark River Slough are not fully understood. Smith and Potts (2002) and Lee et al. (2002) show a slow but steady current of water through the central Bay from the west that likely contains some fraction of Shark River Slough water and nutrients. As noted earlier, Boyer et al. (1999) observed declining average salinity in the central and western Bay with increasing fresh water discharge from Shark Slough. If the currents in 2050 are similar to present patterns, the central and western portions of Florida Bay would be exposed to increased nitrogen fluxes from Everglades restoration, even if water flows remain about the same in Taylor Slough and the eastern Bay.

In conclusion, the ecological response to increased freshwater discharge to Florida Bay seems less certain than it once appeared. Moreover, it seems likely that increased freshwater flow, if it does occur, will increase nutrient loading to the Bay, whether it comes directly into the Bay or indirectly through Shark River Slough. Field and laboratory observations provide circumstantial, but strong, evidence that the response of these marine ecosystems, which historically have been very low in nutrients, to increased nutrient loading will be an increase in phytoplankton blooms and a decrease in water clarity. The effects of such a fundamental water column shift on the seagrasses and associated resources of Florida Bay will be important to resource managers in the region, particularly because it is likely that these changes will be viewed by many as undesirable.

3

Research Needs

Major changes in the Florida Bay ecosystem have been observed in the last two decades, including *Thalassia* die-off (Zieman et al., 1988), declining shrimp harvests (Ehrhardt and Legault, 1999), hypersalinity (McIvor et al., 1994), increased turbidity and decreased light penetration (Boyer et al., 1999), declines in the sponge population (Butler et al., 1995), and others. These changes have led to an increased focus on Florida Bay by the research community. As noted in Chapter 1, this research is being conducted by numerous governmental and academic institutions.

Some, but presumably not all, of these changes may be due to water management practices in south Florida. Since the construction and operation of the Central and South Florida Project, the amount and timing of freshwater runoff into Florida Bay has been radically different from the historical patterns, including a reduction in discharges into the Bay (Light and Dineen, 1994). Although no direct cause-effect relationship has linked south Florida water management to the sea-grass die-off, hypersalinity, declining shrimp harvests, and water-quality changes in the Bay stimulated much of the concern about the condition of the Florida Bay ecosystem reflected in the 1992 Federal Water Resources Development Act, which authorized the Restudy of the Central and South Florida Project.

This chapter summarizes some of the research that needs to be undertaken to resolve questions of the potential effects of the plan that arose out of the Restudy, i.e., the Comprehensive Everglades Restoration Plan (CERP), on Florida Bay. No implication is intended that research conducted to address other science or management goals is unimportant or unnecessary.

This report places emphasis on the potential for changes in the Bay such as increased sediment turbidity and algal blooms resulting from increased freshwater flows from the Everglades. Therefore, this chapter is organized into research needs for (1) the water budget of the Bay, (2) nutrients dissolved or suspended in this water, and (3) modeling approaches to enhance understanding of the connections between the freshwater and marine environments.

Water Budget

Without knowledge of the sensitivity of the Florida Bay ecosystem to alterations in freshwater and nutrient inputs, the effects of the CERP on Florida Bay are difficult to predict. The *overall* water budget (i.e., inflows, outflows, and storage) for the Bay appears to be dominated by rainfall and evaporation (Nuttle et al., 2000). However, these authors recognized the need for improved estimates of precipitation, evaporation, changes in runoff resulting from natural causes and water management, and exchange with the Gulf of Mexico and the Atlantic Ocean. It is even more important, however, to recognize that the overall budget of the Bay is not necessarily relevant to individual sectors of the Bay. For example, salinity and nutrient data clearly show that surface flow from the Everglades is important in the northeastern Bay (Boyer et al., 1999).

Priority research topics concerning water balance include the following:

- Accurate quantification of surface runoff into Florida Bay clearly is a high priority. Ongoing studies (Hittle et al., 2001) have been measuring freshwater discharge into northeastern Florida Bay at five creeks in the Taylor Slough and C-111 canal basins since late 1994. Continued long-term monitoring of spatial and temporal variations in surface runoff will reduce the uncertainty of the effect of the CERP on the Florida Bay water budget, and this activity is recommended. Although sheetflow over the Buttonwood Embankment (Figure 2) between the channels that breach it was not believed to be important by Hittle et al. (2001) or Davis et al., (2002), this should be confirmed by installing instrumented sites along the embankment. Likewise, significant additional freshwater may enter the Bay by diffuse seepage through the embankment, and this process also merits evaluation. Continued research on the hydrodynamic characteristics and net outflow of water from the Shark River Slough basin (e.g., Levesque and Patino, 2001) may be useful, given that some of this water appears to reach the central Bay (Lee et al., 2002).

- Aside from the shallow seepage referred to above, groundwater inputs may be important to the system in other areas. Although nearly all of this influx must be saline (Fitterman and Deszcz-Pan, 2001; Corbett et al., 1999, 2000), saline discharge that has circulated through the phosphorite-rich quartz sand deposits described in Cunningham et al. (1998) may be a significant source of phosphorus in the northwestern Bay (Brand, 2002). Furthermore, a better understanding of how the proposed hydrologic changes in the CERP may affect the position of the freshwater-saltwater contact in the subsurface should be developed. This kind of understanding may be achieved through modeling of the wetland/coastal transition zone, as described later in this chapter.

- The South Florida Water Management Model (SFWMM) predicts about the same annual runoff toward Florida Bay via Craighead Basin and Taylor Slough in the year 2050 as occurs presently (Figure 3). In contrast, flows down Shark River Slough would increase from 702,000 to 1,255,000 acre-ft per year (8.66×10^8 to 1.55×10^9 cubic meters per year) (Figure 4). Hence, on average, the full CERP implementation has a minimal effect on the volume of direct fresh surface water flow into the Bay but a potentially significant effect on the discharge out of Shark River Slough and Whitewater Bay, which eventually reaches the Bay. Additional definition of these flow pathways is urgently needed for modeling of impacts in the Bay. Variability must be accounted for through analysis of the full 31-year simulation period.

- After mixing with seawater, water from Shark River Slough tends to reach Florida Bay in about two to six weeks (Lee et al., 2002) and may still influence the Bay's nutrient budget. The effects on the Bay of a Shark River Slough increase and a stable Taylor Slough/Craighead Basin freshwater flow have not been investigated, however. It is important to remember that the flows shown in Figures 3 and 4 are modeled flows, and may not accurately represent actual flows under current conditions or those after implementation of the CERP. Furthermore, the hydrologic models are subject to change (e.g., eventual conversion from the SFWMM to the South Florida Regional Simulation Model) as well as changing simulation conditions (e.g., use of a 36-year meteorological record vs. a 31-year record). Nonetheless, the implications of this potentially large shift in the magnitude of freshwater discharges down Shark River Slough as a result of the CERP need to be investigated.

- Although surface and groundwater inputs are of most direct use in understanding the effects of the CERP on the Bay, an improved water budget for the Bay as a whole, including natural variability, will help put these inputs into a larger context. Nuttle et al. (2000) estimated a water budget using evaporation rates from a rough salt balance calculated for four regions of Florida Bay, with pan evaporation rates used to distribute annual evaporation seasonally. Price et al. (2001) have been using four approaches (energy flux, vapor flux, stable isotopes, and a box model of salinity) to estimate mean rates of evaporation and its spatial and temporal variation. Both the energy and vapor flux methods depend upon measurement of net radiation, water and air temperature, relative humidity, rainfall, and wind speed and direction. Price et al.'s (2001) study is only a two-year project; continuation of such work (and expansion of the monitoring station network) would provide the CERP with more reliable information about evaporation and precipitation over Florida Bay and about long-term climate changes.

Dissolved Nutrients

Modification of freshwater delivery to the Bay as a result of the CERP has the potential to affect both overall nutrient loadings and the limiting nutrient status of the water by changing the nitrogen:phosphorus ratio in the source waters and ultimately in Florida Bay and the nearby coastal ocean. The consequences of changes in water quality on the Florida Bay ecosystem are a major uncertainty of the CERP. Consequently, studies are needed to address the following questions or concerns:

• Most of the dissolved nitrogen (which is mostly organic nitrogen) in Taylor Slough and the Craighead and C-111 basins passes through mangrove wetlands at the southern boundary of the Everglades. How is the nature of the dissolved nitrogen altered in this environment? Research on this and the following question is ongoing (e.g., Reyes et al., 2001a, b).

• What are the overall loadings of nitrogen and phosphorus from these wetlands through the tidal creeks that pass through the Buttonwood Embankment (Hittle et al., 2001) and into the Bay? How do these vary seasonally?

• The apparent increase in total phosphorus concentrations in Shark River Slough water as it passes through the mangrove zone, as reported by Rudnick et al. (1999), should be examined again. If confirmed, the source of the phosphorus should be determined to evaluate its potential contribution to future enhanced water flow from the CERP.

• One of the most significant uncertainties regarding nitrogen concerns the potential *bioavailability* of the dissolved organic nitrogen and phosphorus (DON and DOP) in Everglades water (including waters in canals that drain south from the Everglades Agricultural Area and Conservation Areas) entering Florida Bay. This includes two issues: (1) what fraction of the DON can be assimilated directly by primary producers (algae and macroscopic aquatic vegetation) in the Bay, and (2) what fraction of the DON can be transformed into nitrogen forms that can be assimilated by algae and macroscopic aquatic plants within the time frame that inflowing water resides in the Bay.

• The chemical nature of DON in Everglades surface water is not known, and such information would be useful to predict the ease and rates of its transformation to inorganic nitrogen forms via microbial metabolism and photochemical degradation. Overall, the factors and processes that control the rates of DON and DOP transformation into bioavailable forms under ambient conditions are poorly understood. What percentage of the DON and DOP pool is refractory within the time constraints imposed by water movement and circulation patterns into and out of the Bay? (Research on internal Bay phosphorus cycling is ongoing by Florida International University scientists.) Much research on Florida Bay is focused on interactions between dissolved, mineralized, sorbed, and gaseous phases of the major limiting nutrients. These transformations can be enormously complex, varying spatially and temporally depending on season, physico-chemical conditions, and a wide range of additional factors. Thus, although process-based work is resource consumptive, it is essential that the continuity of this work be maintained. The key to predicting the effects of various changes to the Everglades on the Bay systems is the use of process-based (mechanistic) models that are based on understanding transformations of the major limiting nutrients.

• Ecosystem response to nitrogen and phosphorus inputs is important in understanding the effects of the CERP on Florida Bay. This is true regardless of the source and/or form (organic vs. inorganic) of the nutrients that enter the Bay. Quantifying the magnitude of nutrient loadings by source (e.g., organic soils oxidation, urban and agricultural runoff, and regional atmospheric deposition) is relevant, however, if steps to reduce nutrient loading to the Bay become necessary. In support of this potential mitigation

effort, an effort should begin now to synthesize and evaluate present knowledge of the sources and transport of nitrogen and phosphorus within the Everglades system.
- The overall nutrient budgets for the Bay are important in order to place the findings of the earlier questions into context.

Historical Characterization of the Bay

Not only is it clear that the response of Florida Bay to the CERP is uncertain, but the historical water quality and ecological conditions of the Bay are at least as uncertain. On the basis of presentations made to the CROGEE at its meetings, a restored Florida Bay appears to imply to the public and to many scientists clear water, presence of sea grass, good fishing, and ample opportunities for other recreational activities such as diving. But to what period of history do these conditions correspond? Knowledge of the Bay's condition many decades to centuries ago is limited, and it would be useful to prepare as accurate a historical timeline as possible based on measured and anecdotal information. In this way managers, the public, and the scientific community may be made aware of the proposed future condition in relation to the recent and more distant past.

Modeling and Hydrodynamics

Numerical models are an essential research and management tool needed for integrating and synthesizing results from most real and hypothesized scenarios in the Everglades. Of particular interest for this report are models that may be used to link the hydrodynamic and ecological response of Florida Bay to changes in Everglades hydrology and to provide decision support and analysis tools to restoration planners.

Some success has been achieved in modeling Everglades hydrology. For example, the South Florida Water Management Model (SFWMM; http://www.sfwmd.gov/org/pld/hsm/models/sfwmm/) has been used since 1984 to simulate the hydrology and management of the water resources system of south Florida, and to evaluate hydrologic and hydraulic restoration options for the Everglades. It covers an area of 7600 square miles using a mesh of two-mile by two-mile cells, and discharges surface flows at the southern end of its area of simulation in the mangrove region of south Florida. A similarly structured model, the Natural System Model (NSM; http://www.sfwmd.gov/org/pld/hsm/models/nsm/), has been used to provide insight into a hypothesized "natural system" condition of the same domain.

However, less success has been achieved on water quality (including salinity) and ecological modeling, which will be required to evaluate the effect of CERP activities on Florida Bay. Typically, a transient, two- or three-dimensional hydrodynamic model is used to "drive" eutrophication and other water quality models that must be developed to evaluate the effects on the Bay of external forcing functions such as freshwater inflows from the Florida peninsula. Such circulation models are complex (Martin and McCutcheon, 1999), and extensive data collection activities and occasional modifications of the numerical code are required to adapt an existing model to a new location. Water quality/ecological models can be equally complex in their kinetic formulations, although their mathematics might not be quite as imposing.

Models for Florida Bay

Several efforts have been made to develop a circulation model for Florida Bay (e.g., Wang and Monjo, 1995; Sheng et al., 1995, 1996; Galperin et al., 1995; Roig, 1996; Davis and Sheng, 1996). While these studies were able to represent currents and salinity in the Bay with some success, their achievement was limited by the lack of long-term data necessary for development of a model that could be used operationally to evaluate management options. Characteristics of such a model would include a demonstration of its ability (through calibration with and comparison to monitoring data) to simulate circulation and salinity in response to forcing functions (e.g., tides, ocean boundary conditions, freshwater inflows, precipitation, evaporation) over a period of years. At present, there is no model that has been adopted for management purposes, nor is there a consensus among agencies about which, if any, of the past studies might represent the most promising starting point for future model development; some of the reasons for this lack of consensus are presented by Hobbie et al. (1999).

The need for such a model is well recognized by Florida Bay and Florida Keys Feasibility Study (FBFKFS) scientists, and is emphasized by Hobbie et al. (2001), who suggest drawing from "community models" for this purpose. Community models are "...constructed by a small team of developers [and] made available to the community of users (with some reasonable restrictions) via downloading from the Internet" (Hobbie et al., 2001). Similarly, the FBFKFS might logically draw upon the "community" of those who have already performed Florida Bay modeling for advice during the model selection phase. It is important to note that some ocean models are not suitable for estuarine simulation because of their fixed orthogonal coordinate systems that cannot resolve the Bay's complex geometry and because of their inability to simulate flooding and drying of tidal flats.

There are two requirements for successfully linking a hydrodynamic model with a water quality model. The first requirement is that the hydrodynamic model satisfy local and global mass conservation over the model grid. The two common numerical options for circulation models are finite difference models and finite element models (Martin and McCutcheon, 1999). The former are typically formulated for a rectangular x-y grid. When formulated using a control volume approach, the mass conservation requirement is guaranteed within a finite difference model. Finite element models commonly use triangular elements of varying size and are easier to adapt to complex geometries, although finite-difference methods may be adapted to curvilinear coordinates (e.g., Chau and Jin, 1995; Sheng et al., 1996) and thus mitigate this relative disadvantage. Finite element models sometimes suffer from local (individual element) problems of mass continuity, and extra effort may be required to formulate the numerical scheme in a manner to ensure local mass balance (Berger and Howington, 2002). Additional "flow projection" techniques may also be applied to finite element model output to ensure local and global mass conservation (Chippada et al., 1998; Riviere and Wheeler, 1999). Indeed, these flow projection methods appear useful for general interpolation of vector flow fields, e.g., from observational data.

The second requirement for linking such models is that the hydrodynamic model information be preserved in the water quality model. This is typically achieved by using the same model grid and time step for both the circulation model and water quality model. Both requirements lead to the conclusion that the same type of model (finite difference or finite element) should be used for both the circulation and quality objectives. In this way, constituent residence times and advective fluxes will be approximated to a higher order of accuracy, leading to concentration of constituents by evaporation, for example. Still another consideration is minimizing execution times for the combined circulation and water-quality model through means such as parallel processing so that long-term water quality simulations may be performed. The FBFKFS is expected to consider all these factors during its review of candidate models.

Notwithstanding the need for improved numerical techniques and hydrodynamic modeling, a key to successful water quality modeling in Florida Bay will be more reliable information on freshwater inflows and the nutrient balance (Berger, R.C. and Dortch, M.S., USACE Waterways Experiment Station, Personal commun., March 2002), as discussed in this report. Flow patterns are tied to salinity, which depends on freshwater inflows. The influence of freshwater nutrient inflows has been emphasized. Hence, circulation and water quality modeling will remain uncertain as long as such boundary influences are uncertain. Cerco et al. (2000) point out several other weaknesses in the knowledge of parameters and coefficients affecting water quality modeling of Florida Bay. In the event that a linked hydrodynamic-ecological model of Florida Bay proves elusive or lengthy in its development, the ability just to simulate salinity and turbidity and to track conservative constituents will still be useful for management purposes.

While waiting for an integrated hydrodynamic-water quality model to be developed, simpler techniques could prove useful. FATHOM (Flux Accounting Tidal Hydrology Ocean Model) is a coarse-grid "box model," capable of long-term simulation (years), but dependent on simplifying assumptions about circulation (Nuttle et al., 2000; Cosby et al., 1999; Cosby et al., in prep.). Hobbie et al. (2001)

indicate that better estimates of inter-box exchange coefficients are now available that could make FATHOM more useful for initial guidance and management purposes and allow it to fill an analytical gap in the interim.

Linking Florida Bay and Everglades Models

If an integrated hydrodynamic-water quality model can be developed for Florida Bay, the challenge will remain to link such a model with the SFWMM or similar model. Such an interface is vital for management purposes such as to study the effect of changes in the magnitude, timing, and spatial distribution of freshwater discharges from the Everglades on the Bay. The coarse, 2-mile by 2-mile grid, large time step, lack of tidal influence, and lack of density-driven subsurface flow features of the SFWMM present challenges to this task. However, at least two efforts are underway to interface the SFWMM (or a similar model) with a Bay model.

The first option is the USGS project "Tides and Inflows in the Mangroves of the Everglades (TIME)" (http://sofia.usgs.gov/projects/time/). This project is developing a linked surface-subsurface water flow model that responds dynamically to freshwater inflows, tides, salinities, rainfall, ET, and related forcing functions (Schaffranek et al., 2001). The goals include coupling with hydrologic modeling at the north end of the simulated region (Tamiami Trail) and quantifying the interrelation of freshwater and salt-water flows at the south end (mangrove region) of the study area. The effect of sea-level rise also can be studied with this model. The model currently is being calibrated against monitored water levels and freshwater fluxes at the Florida Bay interface. Monitoring difficulties include obtaining land elevations on mangrove islands, measuring the bathymetry of mangrove inlets, and defining tidal creek flows and salinities (Hittle et al., 2001). However, TIME has performed well in wetland regions of Everglades National Park where better monitoring data are available. This model probably will not be calibrated and tested sufficiently for use in a management setting until mid-2003, and the agencies have not agreed that this model will serve to bridge the gap between the hydrologic modeling and Florida Bay.

A precursor to the TIME model is the Southern Inland and Coastal Systems (SICS) model (http://sofia.usgs.gov/projects/sheet_flow/), developed for the Taylor Slough and C-111 drainage areas. It already has been used for management decisions regarding the effect of freshwater flows on the salinity of the northeast portion of Florida Bay. The SICS model has the same generic formulation as the TIME model and will be subsumed into the TIME model when the latter is fully operational. Interfacing of the TIME model to the SFWMM at the northern boundary depends primarily on resolution of the 2-mile by 2-mile SFWMM grid with the 500-meter TIME grid and should not be a major problem. However, the 500-meter grid and dynamic nature of the TIME model at the tidal boundary means that it runs on a short (15-minute) time step. Hence, TIME can be used only for episodic events, such as a period of drought or flooding and not for a longer portion of the 31-year SFWMM simulation. On the other hand, most estuarine circulation models are limited to simulating periods of at most several months or a few years duration.

A second modeling option for interfacing the hydrology of the CERP with the hydrodynamics of Florida Bay consists of ongoing model development by the SFWMD of its South Florida Regional Simulation Model (SFRSM) (http://glacier.sfwmd.gov/org/pld/hsm/models/sfrsm/index.html). Using a numerical approach similar to the TIME model, a linked surface-groundwater model is under development that will provide considerable additional spatial and temporal detail in the region south of the Tamiami Trail (Lal, 1998). Eventually, this model may replace the SFWMM over the whole south Florida simulation region, but according to the web site "years of development and testing will be needed before SFRSM becomes fully operational for the entire system." Fortunately, initial implementation of the SFRSM will be in the area of Everglades National Park, the region of concern here. But the USGS, SFWMD, and Corps of Engineers have not reached consensus on how to effect the interfacing of hydrologic modeling and circulation modeling, and both the USGS and SFWMD modeling efforts are proceeding without such a mandate. Both efforts should prove useful to address many questions aside from just the interfacing issue but the three agencies should reach a consensus on the hydrologic model most likely to serve as an interface between the overall hydrologic modeling of the CERP and circulation modeling in the Bay.

Statistical models (e.g., correlations between monitored freshwater flows and Bay salinity and algal conditions, time series analysis) may serve as a temporary bridge between CERP forcing functions and Bay response (Hobbie et al., 2001). Development of inferences from models of this type would be useful while the longer and more expensive effort to develop, calibrate and test Florida Bay circulation and water-quality models and hydrologic models linking the land and the Bay is undertaken.

Other Influences on the Bay

Although the effects of CERP management decisions on the water quality and character of Florida Bay are the primary focus of this report, it is important to consider other influential factors, especially if model predictions prove to be highly inaccurate. For instance, the Bay has been shown to be influenced by runoff from as far away as the Mississippi River - albeit minimally and from a highly unusual hydrologic event–the major flood of 1993 (Ortner et al., 1995; Gilbert et al., 1996). Atmospheric deposition of nutrients derived from distant anthropogenic sources contributes to nutrient loads. Sea level and climate change effects may alter boundaries and bathymetry gradually in the long term while hurricanes could dramatically affect the Bay in the short term. The Florida Keys Tidal Restoration Plan (http://www.evergladesplan.org/pm/projects/proj_31.shtml), a component of the CERP, will restore tidal connections between Florida Bay and the Atlantic Ocean at four locations beneath US Route 1 and the Flagler Railroad causeway and will alter Florida Bay circulation patterns to an unknown extent. Future circulation modeling should help determine the potential effect of this action. The greatest unknowns, however, are probably those associated with the human response to the restoration and to patterns of growth in south Florida. The restoration and a reliable water supply will likely induce and facilitate population growth, thus placing additional, unexpected demands on natural resources such as the Everglades and Florida Bay. How much will such additional use degrade these resources and negate restoration efforts? Additional research and consideration of human factors are needed to illuminate these issues.

Research Management

The management, coordination, and structure of ongoing research in the Bay is generally good. While it is true that some of the issues raised in this report (e.g., the need to establish restoration goals for the Bay) were raised during earlier reviews such as Boesch et al. (1996), knowledge of the Bay has increased greatly since the late 1980s. The extension of the CERP to include the Bay as well as recent activities of the Program Management Committee (PMC) should accelerate progress in addressing these important questions. As described in Chapter 1, several agencies and universities collaborate on research directed toward answers to five central questions developed by the Program Management Committee (PMC). Whether or not research is directed toward these questions and whether or not it leads to synthesis and usefulness for managers are the primary criteria affecting recommendations for new and continuing support from the PMC – an effective strategic plan for research management. All research is peer reviewed, at least through the biannual Florida Bay Science Conferences as it is prepared for peer-reviewed journals. The effectiveness of each conference is evaluated by the Florida Bay Science Oversight Panel (e.g., Hobbie et al., 2001). The oversight and coordination exhibited by these several groups are commendable.

A related issue is one of adequate time to perform the needed research. For instance, the FBFKFS is scheduled for four years, including approximately one year at the beginning devoted mainly to planning and a final year devoted mainly to preparation of documentation of the study. This leaves only about two years to perform the bulk of the research, including complex hydrodynamic and water-

4

Conclusions and Recommendations

Because the Comprehensive Everglades Restoration Plan (CERP) is an ambitious and comprehensive enterprise, with a long time horizon, it is critical that it be responsive to new information as it becomes available from the extensive ongoing research and monitoring programs throughout south Florida. One important assumption often made by scientists and managers associated with the planned restoration of the Everglades, and by the public, is that the increased flows of water deemed necessary to restore the extensive wetland marsh habitats of the Everglades also will contribute to the restoration of Florida Bay. However, the scientific evidence suggesting that the CERP may change the marine environment in ways that are not fully understood, and may be perceived as undesirable, is sufficiently persuasive that the issue should be the subject of a focused technical review and evaluation. This review should be carried out as an early activity within the FBFKFS so that the conclusions can have an influence on early stages of Everglades restoration planning. The results of this analysis need to be evaluated by resource managers and planners of CERP so that appropriate consideration and management actions can be taken.

Major considerations leading to this position are as follows: (1) While it remains debatable how much new freshwater flow will enter the Bay proper because of the CERP, recent physical observations demonstrate that there is commonly a hydraulic connection, albeit with seawater mixing and a time lag, between Shark River Slough discharge and the interior of Florida Bay. (2) Some fraction of the DON (dissolved organic nitrogen) that would accompany increased freshwater flows from the Everglades will likely be available, either directly or indirectly, to support undesirable algal blooms within Florida Bay. (3) Finally, enhanced blooms of phytoplankton and/or macroalgae may reduce seagrass cover and expose sediments to resuspension. Such resuspension will increase turbidity within the Bay and contribute to additional seagrass loss.

Thus, the consequences of the CERP may be a Florida Bay that differs markedly from the "gin clear" bay of the 1960s and 1970s. The CERP goals, if achieved, may instead result in conditions in Florida Bay that are not viewed in a positive light by the public. These conditions may include increased frequency, extent, and duration of phytoplankton blooms, as well as macroalgal blooms, both of which may threaten seagrass distribution.

The importance of this issue has been recognized by some partners in the Florida Bay and CERP research communities, and several projects are underway or will soon begin that will be helpful in addressing it. Research in the following areas is particularly important:

- Some components of the water budget for the Bay are poorly known, but understanding these fluxes is critical for evaluating the water quality in the Bay. In particular, estimates of groundwater discharge to the Bay differ by many orders of magnitude. Although this discharge is saline it may be a significant source of nutrients. Moreover, the higher water levels produced by the CERP may affect the magnitude of these fluxes and, in extreme cases, may result in direct discharge of fresh groundwater to the Bay. Also of importance to both salinity and nutrient fluxes is a full characterization and quantification of quality modeling in Florida Bay. It is important that enough time be available for essential research and that project efforts not be rushed unreasonably at the expense of science.

surface runoff in Taylor Slough, the Craighead Basin, the C-111 canal (Eastern Panhandle), and Shark River Slough, including the seasonality of flow. Diffuse seepage through the Buttonwood Embankment should also be investigated.

- On the basis of modeling, annual freshwater flows through the Taylor Slough/Craighead Basin region to Florida Bay in 2050 will be about the same (127,000 acre-ft per year; 15.7×10^7 m^3/year) under the expanded CERP scenario D13R4 as in the simulated current condition of 126,000 acre-ft per year (15.5×10^7 m^3/year). Thus, this particular hydrologic component of the CERP is unlikely to affect salinity levels in the northeastern Bay. At the same time, annual freshwater flow through Shark River Slough, some of which migrates to the Bay indirectly and with a time lag, is projected to increase by almost 80% under the CERP scenario D13R4. The effects on nitrogen and phosphorus fluxes of increasing this freshwater flux to the Bay's diffuse northwestern boundary need to be quantified.

- Total loads of nitrogen from freshwater sources should be estimated as accurately as possible. A better understanding of the transport, bioavailability, and rates of transformation of DON and DOP (dissolved organic phosphorus) into forms that can be used by algae and macroscopic aquatic plants) of dissolved organic nitrogen is needed to provide insight into the effects of increases in nutrient-bearing freshwater flows to the Bay. Quantifying the magnitude of nutrient loadings by source (e.g., organic soils oxidation, urban and agricultural runoff, and regional atmospheric deposition) also will become relevant if steps to reduce nutrient loading to the Bay become necessary.

- A historical characterization of the Bay's water quality would be very useful for a perspective on restoration goals. Such a characterization would be based on anecdotal as well as any scientific information available.

- There currently is no Florida Bay circulation model suitable for research and management purposes, although there are several candidates; such a model is essential to support a Bay water quality model and thus facilitate analysis of CERP effects on the Bay. The difficult and time-consuming tasks of selection, development, and application of a circulation model and water-quality model for the Bay should be key components of the Florida Bay & Florida Keys Feasibility Study.

- To evaluate the effects of the CERP on Florida Bay, there must be a linkage of the output of the South Florida Water Management Model (SFWMM), which has a southern boundary of the mangrove zone, and input to the Bay models. One possibility for bridging this gap is the USGS TIME model, but its readiness for this purpose probably will not occur until well into 2003. The USGS SICS model (an early version of TIME) already is being used for this purpose in the Taylor Slough/C-111 canal area, although not formally interfaced to any Bay models. Additional questions of temporal and spatial resolution must also be resolved. Another option is the similarly structured SFWMD South Florida Regional Simulation Model (SFRSM), also currently under development. An interagency agreement on which model will most usefully serve as an interface between landside hydrologic and Bay hydrodynamic modeling is needed.

- Estimates of the influence of the CERP on Florida Bay inferred from statistical and time series analysis of existing data and/or use of simpler "box models" such as FATHOM (Flux Accounting Tidal Hydrology Ocean Model) may be usefully employed while awaiting development of full-scale simulation models.

- Human factors such as population growth and economic activity; and environmental events whose drivers are distant from Florida Bay and unrelated to CERP activities, such as hurricanes, flooding of the Mississippi River, atmospheric deposition of nutrients from emission sources remote to Florida Bay or its watershed, and sea-level rise; also may influence local conditions in Florida Bay. Research is needed to better define these potential effects and to integrate the results into predictive ecosystem-response models.

There is much to commend in the overall research effort in Florida Bay, and in the goals of the FBFKFS. But the evidence of potential changes in Florida Bay that will be caused by the CERP is sufficiently persuasive that the process of evaluation should begin as an early part of the FBFKFS. The effort required for all these tasks is daunting. Sufficient time and resources should be made available as part of the FBFKFS for essential research to ensure its success and usefulness for management decisions.

References

Antia, N. J., P. J. Harrison, and L. Oliveriar. 1991. The role of dissolved organic nitrogen in phytoplankton nutrition, cell biology and ecology. Phycologia 30:1-89.

Armentano, T. V., M. Robblee, P. Ortner, N. Thompson, D. Rudnick, and J. Hunt. 1994. Science Plan for Florida Bay. Unpublished report provided to the Interagency Working Group on Florida Bay.

Beardall, J., E. Young, and S. Roberts. 2001. Approaches for determining phytoplankton nutrient limitation. Aquatic Sciences 63:44-69.

Berger, R. C. and S. E. Howington. 2002. Discrete Fluxes and Mass Balance in Finite Elements. Journal Hydraulic Engineering. 128(1):87-92.

Boyer, J. N., J. W. Fourqurean and R. D. Jones. 1997. Spatial characterization of water quality in Florida Bay and Whitewater Bay by multivariate analyses: zones of similar influence. Estuaries 20:743-758.

Boyer, J. N., J. W. Fourqurean, and R. D. Jones. 1999. Seasonal and long-term trends in water quality of Florida Bay (1987-97). Estuaries 22:417-430.

Brand, L. 2002. The transport of terrestrial nutrients to south Florida coastal waters. Pp. 361-413 in The Everglades, Florida Bay, and Coral Reefs of the Florida Keys, J. W. Porter and K. G. Porter, eds. Boca Raton, Fla.: CRC Press.

Burnett, W., J. Chanton, J. Christoff, E. Kontar, S. Krupa, M. Lambert, W. Moore, D. O'Rourke, R. Paulsen, C. Smith, L. Smith, and M. Taniguchi. 2002. Assessing methodologies for measuring groundwater discharge to the ocean. Eos Trans. AGU 83:117, 122-123.

Bush, P. W. and R. H. Johnston. 1988. Ground-water hydraulics, regional flow, and ground-water development of the Floridan Aquifer System in Florida and in parts of Georgia, South Carolina, and Alabama. United States Geological Survey Professional Paper 1403-C. Reston, Va.: U.S. Geological Survey.

Bushaw-Newton, K. L., and M. A. Moran. 1999. Photochemical formation of biologically available nitrogen from dissolved humic substances in coastal marine systems. Aquatic Microbial Ecology 18 (3): 285-292.

Bushaw, K. L., R. G. Zepp, M. A. Tarr, D. Schulz-Jander, R. A. Bourbonniere, R. E. Hodson, W. L. Miller, D. A. Bronk, and M. A. Moran. 1996. Photochemical release of biologically available nitrogen from aquatic dissolved organic matter. Nature 381:404-407.

Butler, M. J., J. H. Hunt, W. F. Hernkind, M. J. Childress, R. Bertelsen, W. Sharp, T. Matthews, J. M. Field, and H. G. Marshall. 1995. Cascading disturbances in Florida Bay, USA: Cyanobacteria blooms, sponge mortality, and implications for juvenile spiny lobsters *Panulirus argus*. Mar. Ecol. Prog. Ser. 129:119-125.

Carlson, P. R., L. A. Yarbro, and T. R. Barber. 1994. Relationship of sediment sulfide to mortality of *Thalassia testudinum* in Florida Bay. Bulletin of Marine Science 54:733-746.

Carlsson, P. and E. Granéli. 1998. Availability of humic bound nitrogen for coastal phytoplankton. Estuarine Coastal Shelf Sci. 36:433-447.

Carlsson, P., E. Granéli, and A. Z. Segatto. 1999. Cycling of biologically available nitrogen in riverine humic substances between marine bacteria, a heterotrophic nanoflagellate and a photosynthetic dinoflagellate. Aquatic Microbial Ecology 18(1):23-26.

Carlsson, P., A. Z. Segatto, and E. Granéli. 1993. Nitrogen bound to humic matter of terrestrial origin - a nitrogen pool for coastal phytoplankton? Marine Ecology Progress Series 97:105-116.

Cerco, C. F., B. W. Bunch, A. M. Teeter, and M. S. Dortch. 2000. Water Quality Model of Florida Bay. ERDC/ELTR-00-10, U. S. Army Engineer Research and Development Center, Vicksburg. Online. Available at http://www.wes.army.mil/el/elpubs/pdf/trel00-10.pdf. Accessed March 2002.

Chau, K. W. and H. .S. Jin. 1995. Numerical solution of two-layered, two-dimensional tidal flow in boundary-fitted orthogonal curvilinear coordinate system. Int. J. Numerical Methods in Fluids 21:1087-1107.

Chippada, S., C. N. Dawson, M. L. Martinez, and M. F. Wheeler. 1998. A projection method for constructing a mass conservative velocity field. Computer Methods in Applied Mechanics and Engineering, 157:1-10.

Comprehensive Everglades Restoration Plan (CERP). 2001. Project Management Plan: Florida Bay and Florida Keys Feasibility Study (third draft) Online. Available at http://www.evergladesplan.org/pm/program/program_docs/pmp_study_florida/cerp_fb_fk.pdf. Accessed January 30, 2002.

Corbett, D. R., J. Chanton, W. Burnett, K. Dillon, C. Rutkowski, and J. W. Fourqurean. 1999. Patterns of groundwater discharge into Florida Bay. Limnol. Oceanogr. 44:1045-1055.

Corbett, D. R., K. Dillon, W. Burnett, and J. Chanton. 2000. Estimating the groundwater contribution into Florida Bay via natural tracers, ^{222}Rn and CH_4. Limnol. Oceanogr. 45:1546-1557.

Cosby, B. J., Nuttle, W. K., and J. W. Fourqurean. 1999. FATHOM: Model description and initial application to Florida Bay, project completion report. National Park Service, U.S. Dept. of Interior, Washington, D.C.

Cosby, B. J., W. K. Nuttle, and J. W. Fourqurean. In prep. A hydrologic model of Florida Bay based on basin water balances and frictional flows over shoals. For submission to Water Resources Research.

Cunningham, K. J., D. F. McNeill, L. A. Guertin, T. M. Scott, P. F. Ciesielski, and L. de Verteuil. 1998. New Tertiary stratigraphy for the Florida Keys and southern peninsula of Florida. Geological Society of America Bulletin 110:231-258.

D'Sa, E. J., J. B. Zaitzeff, C. S. Yentsch, J. L. Miller and R. Ives. 2002. Rapid remote assessments of salinity and ocean color in Florida Bay. Pp. 451-459 in The Everglades, Florida Bay, and Coral Reefs of the Florida Keys, J. W. Porter and K. G. Porter, eds. Boca Raton, Fla.: CRC Press.

Davis, J. R., and Y. P. Sheng. 1996. Modeling Florida Bay Circulation. Presented at the Second Scientific Symposium on Florida Bay, Florida Sea Grant, Key Largo, Florida, December 1996.

Davis III, S. E., D. L. Childers, J. Cable, J. W. Day Jr., D. T. Rudnick, and F. H. Sklar. 2002. Event driven nutrient dynamics in a Southern Everglades mangrove creek. EOS. Trans. AGU 83(19), Spring Meet. Suppl., Abstract H41B-06.

Davis, S. M., L. H. Gunderson, W. A. Park, J. R. Richardson, and J. E. Mattson. 1994. Landscape dimensions, composition, and function in a changing Everglades ecosystem. Pp. 419-444 in Everglades – The Ecosystem and its Restoration, S. M. Davis and J. C. Ogden, eds. Delray Beach, Fla.: St. Lucie Press.

Durako, M. D. and K. M. Kuss. 1994. Effects of *Labyrinthula* infection on the photosynthetic capacity of *Thalassia testudinum*. Bulletin of Marine Science 54:727-732.

Ehrhardt, N. M. and C. M. Legault. 1999. Pink Shrimp, *Farfantepenaeus duorarum*, Recruitment Variability as an Indicator of Florida Bay Dynamics. Estuaries 22(2B):471-483.

Fitterman D. V. and M. Deszcz-Pan. 2001. Saltwater Intrusion in Everglades National Park, Florida Measured by Airborne Electromagnetic Surveys. First International Conference on Saltwater Intrusion and Coastal Aquifers: Monitoring, Modeling, and Management. Essaouira, Morocco,

April 23 —25, 2001. Rabat, Morocco: Laboratoire d'Analyse des Systèmes Hydrauliques. CD-ROM available at http://www.olemiss.edu/sciencenet/saltnet/.

Florida Coastal Everglades Long Term Ecological Research. 2002. Florida Coastal Everglades LTER Sites. Online. Florida International University. Available at http://fcelter.fiu.edu/overview/FCEslides/pages/Slide29.htm. Accessed August 2002.

Fourqurean, J. W. and M. B. Robblee. 1999. Florida Bay: A history of recent ecological changes. Estuaries 22:345-357.

Fourqurean, J. W., J. C. Zieman, and G. V. N. Powell. 1992. Phosphorus limitation of primary production in Florida Bay: Evidence from the C:N:P ratios of the dominant seagrass *Thalassia Testudinum*. Limnology and Oceanography 37:162-171.

Galperin, B., M. Haines and M. Luther. 1995. The Design of a Modeling Strategy for Florida Bay. Report to Everglades National Park, National Park Service. Tampa:University of South Florida.

Gilbert, P. S., T. N. Lee, and G. Podesta. 1996. Transport of Anomalous Low-Salinity Waters from the Mississippi River Flood of 1993 to the Straits of Florida. Cont. Shelf Res. 16(8):1065-1085.

Gregg, W. H. 1902. Where, When, and How to Catch Fish on the East Coast of Florida. Buffalo: The Matthews-Northrup Works.

Halley, R. B. 2002. Florida Bay's Murky Past. Florida Bay Watch Header. Online. Available at http://sofia.usgs.gov/publications/reports/flbay_murky/. Accessed April 1, 2002.

Halley, R. B., and L. M. Roulier. 1999. Reconstructing the history of eastern and central Florida Bay using mollusk-shell isotope records. Estuaries 22:358-368.

Hittle, C. 2001. Quantity, timing, and distribution of freshwater flows into northeastern Florida Bay. Pp. 11-12 in Program and Abstracts, 2001 Florida Bay Science Conference, Key Largo, Fla., April, 2001. Gainesville: University of Florida Office of Conferences and Institutes. Available at http://sofia.usgs.gov/projects/freshwtr_flow/frshwtrflowabfb2001.html.

Hittle, C., E. Patino, and M. Zucker. 2001. Freshwater flow from estuarine creeks into northeastern Florida Bay: U.S. Geological Survey Water-Resources Investigations Report 01-4164, Reston, Va.: U.S. Geological Survey.

Hobbie, J. E., W. C. Boicourt, L. Deegan, K. L. Heck, Jr., S. C. McCutcheon, J. D. Milliman and H. W. Paerl. 1999. Report of the Florida Bay Science Oversight Panel, Perspectives from the 1999 Florida Bay Science Conference. Program Management Committee of the Interagency Florida Bay Science Program. (Available at http://www.aoml.noaa.gov/ocd/sferpm/oversight_report00.html.)

Hobbie, J. E., W. C. Boicourt, K. L. Heck, Jr., E. T. Houde, S. C. McCutcheon and J. Pennock. 2001. Report of the Florida Bay Science Oversight Panel, Perspectives from the 2001 Florida Bay Science Conference. Program Management Committee of the Interagency Florida Bay Science Program. (Available at http://www.aoml.noaa.gov/ocd/sferpm/oversight_report01.html).

Jackson, J. B. C., M. X. Kirby, W. H. Berger, K. A. Bjorndal, L. W. Botsford, B. J. Bourque, R. H. Bradbury, R. Cooke, J. Erlandson, J. A. Estes, T. P. Hughes, S. Kidwell, C. B. Lange, H. S. Lenihan, J. M. Pandolfi, C. H. Peterson, R. S. Steneck, M. J. Tegner and R. R. Warner. 2001. Historical overfishing and the recent collapse of coastal ecosystems. Science 293:629-637.

Jensen, H. S., K. J. McGlathery, R. Marino, and R. W. Howarth. 1998. Forms and availability of sediment phosphorus in carbonate sand of Bermuda seagrass beds. Limnology and Oceanography 43:799-810.

Kaufman, J. 1994. Numerical models of fluid flow in carbonate platforms: Implications for dolomitization. J. Sed. Res. A64:128-139.

Koopmans, D. J., and D. A. Bronk. 2002. Photochemical production of dissolved inorganic nitrogen and primary amines from dissolved organic nitrogen in waters of two estuaries and adjacent surficial groundwaters. Aquatic Microbial Ecology 26: 295-304.

Lal, W. A. M. 1998. Simulation of Overland and Groundwater Flow in the Everglades National Park. Pp. 610-615 in Proceedings of the International Water Resources Engineering Conference, August 3-7, Memphis, Tenn. Reston, Va.: American Society of Civil Engineers.

Lapointe, B. E. and M W. Clark. 1992. Nutrient inputs from the watershed and coastal eutrophication in the Florida Bay Keys. Estuaries 15:465-476.

Lapointe, B. E., W. R. Matzie, and P. J. Barile. 2002. Biotic phase-shifts in Florida Bay and fore reef communities of the Florida Keys: Linkages with historical freshwater flows and nitrogen loading from Everglades runoff, Pp. 629-648 in The Everglades, Florida Bay, and Coral Reefs of the Florida Keys, J. W. Porter and K. G. Porter, eds. Boca Raton, Fla.: CRC Press.

Lee, T. N., E. Williams, E. Johns, D. Wilson, and N. P. Smith. 2002. Transport processes linking south Florida coastal ecosystems. Pp. 309-342 in The Everglades, Florida Bay, and Coral Reefs of the Florida Keys, J. W. Porter and K. G. Porter, eds. Boca Raton, Fla.: CRC Press.

Levesque, V. A. and E. Patino. 2001. Hydrodynamic characteristics of estuarine rivers along the southwestern coast of Everglades National Park. Pp. 23-25 in Program and Abstracts, 2001 Florida Bay Science Conference, Key Largo, Fla., April, 2001. Gainesville: University of Florida Office of Conferences and Institutes.

Li, L., D. A. Barry, E. Stagnitti, and J. Y. Parlange. 1999. Submarine groundwater discharge and associated chemical input to a coastal sea. Water Res. Research 35:3253-3259.

Martin, J. L. and S. C. McCutcheon. 1999. Hydrodynamics and Transport for Water Quality Modeling. New York: Lewis Publishers.

Maybeck, M. 1982. Carbon, nitrogen, and phosphorus transport by world rivers. American J. Sci. 282:401-450.

McCarthy, M., T. Pratum, J. Hedges, and R. Benner. 1997. Chemical composition of dissolved organic nitrogen in the ocean. Nature 390:150-154.

McIvor, C. C., J. A. Ley, and R. D. Bjork. 1994. Changes in freshwater inflow from the Everglades to Florida Bay including effects on biota and biotic processes: a review. Pp. 117-146 in Everglades: The Ecosystem and Its Restoration, S. M. Davis and J. C. Ogden, eds. Delray Beach, Fla.: St. Lucie Press.

McPherson, B. F., and R. Halley. 1996. The south Florida environment – A region under stress. U.S. Geological Survey Circular 1134. Reston, Va.: U.S. Geological Survey.

Light, S. S., and J. W. Dineen. 1994. Water control in the Everglades: A historical perspective. Pp. 47-84 in Everglades – The Ecosystem and its Restoration, S. M. Davis and J. C. Ogden, eds. Delray Beach, Fla.: St. Lucie Press.

Nuttle, W. K., J. W. Fourqurean, B. J. Cosby, J. C. Zieman, and M. B. Robblee. 2000. Influence of net freshwater supply on salinity in Florida Bay. Water Resources Research 36:1805-1822.

Ortner, P., T. N. Lee, P. Milne, R. Zika, M. E. Clarke, G. P. Podesta, P. Swart, P. A. Tester, L. P. Atkinson and W. Johnson. 1995. Mississippi River flood waters that reached the Gulf Stream. J. Geophys. Res., 100(C7):13595-13601.

Pardo, L. H., C. T. Driscoll, and G. E. Likens. 1995. Patterns of nitrate loss from a chronosequence of clear-cut watersheds. Water Air Soil Pollution 85:1659-1664.

Patino, E. and C. Hittle. 2000. Magnitude and distribution of flows into Northeastern Florida Bay. U.S. Department of the Interior, USGS Fact Sheet FS-030-00. Washington, D.C.: USGS.

Prager, E. J. and R. B. Halley. 1999. The Influence of Seagrass on Shell Layers and Florida Bay Mudbanks. Journal of Coastal Research 15:1151-1162.

Price, R. M., P. K. Swart, and W. K. Nuttle. 2001. Estimating evaporation rates in Florida Bay. Pp. 29-30 in Program and Abstracts, 2001 Florida Bay Science Conference, Key Largo, Fla., April, 2001. Gainesville: University of Florida Office of Conferences and Institutes.

Reyes, E., J. Cable, J. W. Day, D. Rudnick, F. Sklar, C. Madden, S. Kelly, C. Coronado-Molina, S. Davis and D. Childers. 2001a. Nutrient Dynamics in the Mangrove Wetlands of the Southern Everglades – 5 Year Project Overview. Pp. 93-95 in Program and Abstracts, 2001 Florida Bay Science Conference, Key Largo, Fla., April, 2001. Gainesville: University of Florida Office of Conferences and Institutes.

Reyes, E., J. W. Day, S. Davis, and C. Coronado-Molina. 2001b. Nutrient Dynamics and Exchange within a Mangrove Creek and Adjacent Wetlands in the Southern Everglades. Pp. 91-92 in

Program and Abstracts, 2001 Florida Bay Science Conference, Key Largo, Fla., April, 2001. Gainesville: University of Florida Office of Conferences and Institutes.

Riviere, B. and M. F. Wheeler. 1999. Locally Conservative Algorithms for Flow. Proceedings of Mathematics of Finite Elements and Applications MAFELAP 1999, ed. J. Whiteman, pp. 29-46, Elsevier, Amsterdam.

Robblee, M. B., T. R. Barber, P. R. Carlson, M. J. Durako, J. W. Fourqurean, L. K. Muehlstein, D. Porter, L. A. Yarbro, R. T. Zieman, and J. C. Zieman. 1991. Mass mortality of the tropical seagrass *Thalassia testudinum* in Florida Bay (USA). Mar. Ecol. Prog. Ser. 71:297-299.

Roig, L.C. 1996. CEWES Hydrodynamic Modeling of Florida Bay. Overheads, Florida Bay Interagency Modeling Workshop, Marathon, Fla., April, 1996.

Rudnick, D. T., Z. Chen, D. L. Childers, J. N. Boyer, and T. D. Fontaine, III. 1999. Phosphorus and nitrogen inputs to Florida Bay: The importance of the Everglades watershed. Estuaries 22:398-416.

Sanford, W. W., F. F. Whitaker, P. L. Smart, and G. Jones. 1998. Numerical analysis of seawater circulation in carbonate platforms: I. Geothermal convection. Am. J. Science 298:801-828.

Schaffranek, R. W., J. L. Jenter, C. D. Langevin and E. D. Swain. 2001. The tides and inflows in the mangroves of the Everglades project. Pp. 37-39 in Program and Abstracts, 2001 Florida Bay Science Conference, Key Largo, Fla., April, 2001. Gainesville: University of Florida Office of Conferences and Institutes.

Sheng, Y. P., J. R. Davis, and Y. Liu. 1995. A Preliminary Modeling Study on Circulation and Transport in Florida Bay. Final Report to the Everglades National Park and Dry Tortugas National Park, National Park Service. Gainesville: Coastal & Oceanographic Engineering Department, University of Florida.

Sheng, Y. P., J. R. Davis and Y. Liu. 1996. Circulation and Transport in Hypersaline Florida Bay. Proceedings of the 25th International Conference on Coastal Engineering. ASCE.

Seitzinger, S. P. and R. W. Sanders. 1997. Contribution of dissolved organic nitrogen from rivers to estuarine eutrophication. Marine Ecology Progress Series 159:1-12.

Seitzinger, S. P., R. W. Sanders, and R. Styles. 2002. Bioavailability of DON from natural and anthropogenic sources to estuarine plankton. Limnol. Oceanogr., 47:353–366.

Sklar, F. H. and J. A. Browder. 1998. Coastal environmental impacts brought about by alterations to freshwater flow in the Gulf of Mexico. Environmental Management 22, 547-562.

Sklar, F., C. McVoy, R. VanZee, D. E. Gawlik, K. Tarboton, D. Rudnick, and S. Miao. 2002. The effects of altered hydrology on the ecology of the Everglades. Pp. 39-82 in The Everglades, Florida Bay, and Coral Reefs of the Florida Keys, J. W. Porter and K. G. Porter, eds. Boca Raton, Fla.: CRC Press.

Smith, N. P. and Pitts. 2002. Regional-scale and long-term transport patterns in the Florida Keys. Pp. 343-360 in The Everglades, Florida Bay, and Coral Reefs of the Florida Keys, J. W. Porter and K. G. Porter, eds. Boca Raton, Fla.: CRC Press.

South Florida Water Management District (SFWMD). 1998. Everglades Annual Report. West Palm Beach, Fla.: SFWMD.

Stepanauskas, R., L. Leonardson, and L. J. Tranvik. 1999. Bioavailability of wetland-derived DON to freshwater and marine bacterioplankton. Limnology and Oceanography 44(6):1477-1485.

Tarr, M. A., W. W. Wang, T. S. Bianchi, and E. Engelhaupt. 2001. Mechanisms of ammonia and amino acid photoproduction from aquatic humic and colloidal matter. Water Research 35:3688-3696.

Tomas, C. R., B. Bendis, and K. Johns. 1999. Role of nutrients in regulating plankton blooms in Florida Bay. Pp. 323-337 in The Gulf of Mexico Large Marine Ecosystem: Assessment, Sustainability, and Management, H. Kumpf, K. Steidinger, and K. Sherman, eds. Malden, Mass.: Blackwell Science.

Top, Z., L. E. Brand, R. D. Corbett, W. Burnett, and J. Chanton. 2001. Helium and radon as tracers of groundwater input into Florida Bay. J. Coastal Research 17: 859-868.

U.S. Army Corps of Engineers (USACE). 2002. Cape Sable Seaside Sparrow Reasonable and Prudent Alternatives Modeling; South Florida Water Management Model, Regional Hydrologic Performance Measures. Online. USACE Jacksonville District. Available at http://hpm.saj.usace.army.mil/csssweb/frame2/pmg/enp/enp.htm. Accessed August 2002.

U.S. Army Corps of Engineers (USACE). 1999. Central and Southern Florida Project Comprehensive Review Study, Final Integrated Feasibility Report and Programmatic Environmental Impact Statement. Jacksonville: USACE.

Wang, J. D. and C. Monjo. 1995. A Study to Define Model and Data Needs for Florida Bay. Technical Report, Applied Marine Physics, Rosenstiel School of Marine and Atmospheric Science, University of Miami, Miami, Fla.

Wiegner, T. N., and S. P. Seitzinger. 2001. Photochemical and microbial degradation of external dissolved organic matter inputs to rivers. Aquatic Microbial Ecology 24: 27-40.

Wright, A. L., and K. R. Reddy. 2001. Phosphorus loading effects on extracellular enzyme activity in Everglades wetland soils. Soil Science Society of America Journal 65:588-595.

Zieman, J. C., J. W. Fourqurean, and R. L. Iverson. 1988. Distribution, abundance and productivity of seagrasses and macroalgae in Florida Bay. Bull. Mar. Sci. 44:292-311.

Zieman, J. C., J. W. Fourqurean and T. A. Frankovich. 1999. Seagrass die-off in Florida Bay: Long-term trends in abundance and growth of turtle grass, *Thalassia testudinum*. Estuaries 22:460-470.

Appendix A
Acronym List

CERP	Comprehensive Everglades Restoration Plan
CROGEE	Committee on Restoration of the Greater Everglades Ecosystem
DIN	Dissolved Inorganic Nitrogen
DIP	Dissolved Reactive Phosphate
DON	Dissolved Organic Nitrogen
DOP	Dissolved Organic Phosphorus
ENP	Everglades National Park
FATHOM	Flux Accounting Tidal Hydrology Ocean Model
FBFKFS	Florida Bay and Florida Keys Feasibility Study
FCE LTER	Florida Coastal Everglades Long Term Ecological Research program
PMC	Program Management Committee of the Florida Bay and Adjacent Marine Systems Science Program
SFRSM	South Florida Regional Simulation Model
SFWMD	South Florida Water Management District
SFWMM	South Florida Water Management Model
SICS	Southern Inland and Coastal Systems
TDN	Total Dissolved Nitrogen
TIME	Tides and Inflows in the Mangroves of the Everglades

Appendix B

Definitions of Model Runs of the South Florida Water Management Model (SFWMM) and Natural System Model (NSM)

1995 Base (95 B or 95BSR in Figures 3 and 4) is a simulation of the South Florida Water Management system that represents the system infrastructure and operations as they were around 1995. The simulation is made using the South Florida Water Management Model (SFWMM). Water demands and land use are representative of those in 1995. The topography used was the best available at that time. A 31-year climatic record (1965-1995) was used in the simulation. The 1995 Base is also referred to as the current condition, or existing condition.

2050 Base (50BSR in Figure 4) is a simulation of the South Florida Water Management system that represents the likely system infrastructure and operations as they would be around 2050 without any of the Comprehensive Everglades Restoration Plan (CERP) projects in place. Non-CERP Projects assumed to be complete by 2050 include Kissimmee River Restoration, the Everglades Construction Project, Herbert Hoover Dyke improvements, the Modified Water Delivery to Everglades National Park and C-111 projects, and several critical projects in the Lower East Coast area. In the SFWMM simulation of the 2050 Base projected (2050) land use and water demands were used with the same 31-year climatic record (1965-1991) as in the 1995 Base. The 2050 Base is also referred to as the future without project condition or no action alternative.

D13R is the simulation of the South Florida Water Management system with completed Comprehensive Everglades Restoration Plan (CERP) projects in 2050, together with non-CERP projects as simulated in the 2050 Base. In the SFWMM simulation of D13R, the same projected land use and water demands as in the 2050 Base were used. Again the 31-year climatic record (1965-1991) was used. Benefits of D13R are determined by comparison with the current condition (1995 Base) and future without project condition (2050 Base). **D13R4** (see Figures 3 and 4) is a scenario based on D13R that would enhance CERP performance by capturing additional water "lost" to tide.

NSM is the Natural System Model simulation that represents the hydrologic response to the pre-drainage Everglades using the same climatic inputs (1965-1995) as were used for the SFWMM. The NSM does not simulate the influences of any man-made features and uses estimates of pre-subsidence topography and historical vegetation cover. The Natural System Model has evolved as knowledge of the pre-drainage everglades has improved. The version of the NSM used to develop the CERP was NSM version 4.5F (NSM in Figure 3 and NSM45F in Figure 4). In the development of the CERP, the NSM pre-drained hydrologic response of the system was, in many cases, used as a target for hydrologic restoration under the assumption that restoration of the hydrologic response that existed prior to drainage of the system would lead to restoration of natural habitats, biota and species. . But whereas NSM water depths are typically used as CERP targets, NSM flow estimates are not necessarily used as targets because of high sensitivity of flow estimates to small changes in water depth estimates.

Source: Ken Tarboton, SFWMD, written commun., April 2002 and Richard Punnett, USACE, personal commun., July 2002. Additional information may be found at http://www.sfwmd.gov/org/pld/restudy/hpm/.

Appendix C

Biographical Sketches of Members of the Committee on Restoration of the Greater Everglades Ecosystem

JEAN M. BAHR, CHAIR, is professor the Department of Geology and Geophysics at the University of Wisconsin-Madison where she has been a faculty member since 1987. She served as chair of the Water Resources Management Program, UW Institute for Environmental Studies, from 1995-99 and she is also a member of the Geological Engineering Program faculty. Her current research focuses on the interactions between physical and chemical processes that control mass transport in ground water. She earned a B.A in geology from Yale University and M.S. and Ph.D. degrees in applied earth sciences (hydrogeology) from Stanford University. She has served as a member of the National Research Council's Board on Radioactive Waste Management and several of its committees.

SCOTT W. NIXON, VICE-CHAIR, is professor of oceanography at the University of Rhode Island. He currently teaches both graduate and undergraduate classes in oceanography and ecology. His current research interests include coastal ecology, with emphasis on estuaries, lagoons, and wetlands. He has served on three National Research Council committees including, most recently, the Committee on Coastal Oceans. Dr. Nixon received a B.A. in biology from the University of Delaware and a Ph.D. in botany/ecology from the University of North Carolina-Chapel Hill.

JOHN S. ADAMS is professor and chair of the Department of Geography at the University of Minnesota. He researches issues relating to North American cities, urban housing markets and housing policy, and regional economic development in the United States and the former Soviet Union. He has been a National Science Foundation Research Fellow at the Institute of Urban and Regional Development, University of California at Berkeley, and economic geographer in residence at the Bank of America World Headquarters in San Francisco. He was senior Fulbright Lecturer at the Institute for Raumordnung at the Economic University in Vienna and was on the geography faculty of Moscow State University. He has taught at Pennsylvania State University, the University of Washington, and the U.S. Military Academy at West Point. His most recent book, Minneapolis-St. Paul: People, Place, and Public Life, looks at the region's growth and at what factors may affect the metropolitan area's future. Adams holds two degree in economics and a doctorate in urban geography from the University of Minnesota.

LINDA K. BLUM is research associate professor in the Department of Environmental Sciences at the University of Virginia. Her current research projects include study of mechanisms controlling bacterial community abundance, productivity, and structure in tidal marsh creeks; impacts of microbial processes on water quality; organic matter accretion in salt marsh sediments; and rhizosphere effects on organic matter decay in anaerobic sediments. Dr. Blum earned a B.S. and M.S. in forestry from Michigan Technological University and a Ph.D. in soil science from Cornell University.

PATRICK L. BREZONIK is professor of environmental engineering and director of the Water Resources Research Center at the University of Minnesota. Prior to his appointment at the University of Minnesota in the mid-1980s, Dr. Brezonik was professor of water chemistry and environmental science at the University of Florida. His research interests focus on biogeochemical processes in aquatic systems, with special emphasis on the impacts of human activity on water quality and element cycles in lakes. He has served as a member of the National Research Council's Water Science and Technology Board and as a member of several of its committees. He earned a B.S. in chemistry from Marquette University and a M.S. and Ph.D. in water chemistry from the University of Wisconsin-Madison.

FRANK W. DAVIS is a Professor at the University of California Santa Barbara (USCB) with appointments in the Donald Bren School of Environmental Science and Management and the Department of Geography. He received his B.A. in biology from Williams College and a Ph.D. from the Department of Geography and Environmental Engineering at The Johns Hopkins University. He joined the Department of Geography at UCSB in 1983, and established the UCSB Biogeography Lab in 1991. His research focuses on the ecology and management of California chaparral and oak woodlands, landscape ecology, regional conservation planning, and spatial decision support systems. He was Deputy Director of the National Center for Ecological Analysis and Synthesis between 1995 and 1998, and currently directs the Sierra Nevada Network for Education and Research Page. Dr. Davis has been a member of three prior NRC committees.

WAYNE C. HUBER is professor and head of the Department of Civil, Construction, and Environmental Engineering at Oregon State University. Prior to moving to Oregon State in 1991, he served 23 years on the faculty of the Department of Environmental Engineering Sciences at the University of Florida where he engaged in several studies involving the hydrology and water quality of south Florida regions. His technical interests are principally in the areas of surface hydrology, stormwater management, nonpoint source pollution, and transport processes related to water quality. He is one of the original authors of the Environmental Protection Agency's Storm Water Management Model (SWMM) and continues to maintain the model for the EPA. Dr. Huber holds a B.S. in engineering from the California Institute of Technology and an M.S. and Ph.D. in civil engineering from the Massachusetts Institute of Technology. He is currently a member of the NRC's Committee on Causes and Management of Coastal Eutrophication.

STEPHEN R. HUMPHREY is dean of the College of Natural Resources and Environment at the University of Florida where he also serves as affiliate professor of Latin American studies, wildlife ecology, and zoology. He also has been the curator in ecology for the Florida Museum of Natural History since 1980. Dr. Humphrey has authored and co-authored numerous articles and books on the effects of urbanization on wildlife. He holds B.A. in biology from Earlham College in Richmond, Indiana and a Ph.D. in zoology from Oklahoma State University. He is former chair of the Environmental Regulatory Commission of the Florida Department of Environmental Regulation and a member of the Florida Panther Technical Advisory Council of the Florida Game Commission.

DANIEL P. LOUCKS is professor of civil and environmental engineering at Cornell University. His research, teaching, and consulting interests are in the application of economics, engineering, and systems theory to problems involving environmental and water resources development and management. Dr. Loucks has taught at a number of universities in the United States and abroad and has worked for the World Bank, and the International Institute for Applied Systems Analysis. He also served as a consultant to a variety of government and international organizations concerned with resource development and management. He is a member of the National Academy of Engineering and is currently a member of the National Research Council's Committee on Risk-Based Analyses for Flood Damage Reduction Studies.

Appendix C

KENNETH W. POTTER is professor of civil and environmental engineering at the University of Wisconsin-Madison. His expertise is in hydrology and water resources, including hydrologic modeling, estimation of hydrologic risk, estimation of hydrologic budgets, watershed monitoring and assessment, and aquatic ecosystem restoration. He received his B.S. in geology from Louisiana State University and his Ph.D. in geography and environmental engineering from The Johns Hopkins University. He has served as a member of the NRC's Water Science and Technology Board and several of its committees.

LARRY ROBINSON is director of the Environmental Sciences Institute at Florida A&M University where he is also a professor. At Florida A&M University he has led efforts to establish B.S. and Ph.D. programs in environmental science in 1998 and 1999, respectively. His research interests include environmental chemistry and the application of nuclear methods to detect trace elements in environmental matrices and environmental policy and management. Previously he was group leader of a neutron activation analysis laboratory at Oak Ridge National Laboratory (ORNL). At ORNL he served on the National Laboratory Diversity Council and was President of the Oak Ridge Branch of the NAACP. Dr. Robinson earned a B.S. in chemistry, summa cum laude, from Memphis State University and a Ph.D. in nuclear chemistry from Washington University in St. Louis, Missouri.

REBECCA R. SHARITZ is professor of botany at the University of Georgia and senior scientist at the Savannah River Ecology Laboratory in Aiken, South Carolina, where she has been the Head of the Division of Wetlands Ecology. Her research focuses on ecological processes in wetlands, including factors affecting the structure and function of bottomland hardwood and swamp forest ecosystems, responses of wetland communities to environmental disturbances, and effects of land management practices on nearby wetland systems. Dr. Sharitz has served on several NRC committees including, The Committee on Restoration of Aquatic Ecosystems: Science, Technology and Public Policy. She received a B.S. in biology from Roanoke College and a Ph.D. in botany and plant ecology from the University of North Carolina.

HENRY J. VAUX, JR. is professor of resource economics at the University of California, Riverside. He currently serves as Associate Vice President - Agricultural and Natural Resource Programs for the University of California system. He previously served as Director of the University of California Water Resource Center. His principal research interests are the economics of water use and water quality. Prior to joining the University of California he worked at the Office of Management and Budget and served on the staff of the National Water Commission. He received a Ph.D. in economics from the University of Michigan in 1973. He recently served as chair of the National Research Council's Water Science and Technology Board.

JOHN VECCHIOLI retired as a hydrologist with the U.S. Geological Survey's Water Resources Division in Tallahassee, Florida and as chief of the Florida District Program. Previously, he was responsible for quality assurance of all technical aspects of ground water programs in Florida. His research interests have included study of hydraulic and geochemical aspects of waste injection in Florida and of artificial recharge in Long Island, N.Y. He has also done research on ground water-surface water interactions in New Jersey and Florida. Mr. Vecchioli received his B.S. and M.S. in geology from Rutgers University. Mr. Vecchioli previously served on the NRC's Committee on Ground Water Recharge.

JEFFREY R. WALTERS is Bailey Professor of Biology at Virginia Polytechnic Institute and State University, a position he has held since 1994. His professional experience includes assistant, associate, and full professorships at North Carolina State University from 1980 until 1994. Dr. Walters has done extensive research and published many articles on the red-cockaded woodpeckers in North Carolina and Florida and he chaired an American Ornithologists Union Conservation Committee Review that looked at the biology, status, and management of the Cape Sable Seaside Sparrow, a bird native to the Everglades. He is a fellow of the American Ornithologist Union, a member of Sigma Xi, American Society of Naturalists, Animal Behavior Society, Audubon Society, Cooper Ornithological Society, Ecological Society of America, Phi Beta Kappa, and many other scientific organizations. His research interests are in cooperative breeding in birds; reproductive biology of precocial birds; primate intragroup social behavior; evolution of cooperative breeding in birds; ecological basis of sensitivity to habitat fragmentation; kinship effects on behavior; and parental behavior on precocial birds. He holds a B.A. from West Virginia University and a Ph.D. from the University of Chicago.